Internet of Things

Technology, Communications and Computing

Series Editors

Giancarlo Fortino, Rende (CS), Italy
Antonio Liotta, Edinburgh Napier University, School of Computing,
Edinburgh, UK

The series Internet of Things - Technologies, Communications and Computing publishes new developments and advances in the various areas of the different facets of the Internet of Things.

The intent is to cover technology (smart devices, wireless sensors, systems), communications (networks and protocols) and computing (theory, middleware and applications) of the Internet of Things, as embedded in the fields of engineering, computer science, life sciences, as well as the methodologies behind them. The series contains monographs, lecture notes and edited volumes in the Internet of Things research and development area, spanning the areas of wireless sensor networks, autonomic networking, network protocol, agent-based computing, artificial intelligence, self organizing systems, multi-sensor data fusion, smart objects, and hybrid intelligent systems.

** Indexing:*Internet of Things* is covered by Scopus and Ei-Compendex **

More information about this series at http://www.springer.com/series/11636

D. Jude Hemanth • J. Anitha
George A. Tsihrintzis
Editors

Internet of Medical Things

Remote Healthcare Systems and Applications

 Springer

Editors
D. Jude Hemanth
Department of Electronics &
Communication Engineering
Karunya Institute of Technology and
Sciences
Coimbatore, Tamil Nadu, India

J. Anitha
Department of Electronics &
Communication Engineering
Karunya Institute of Technology and
Sciences
Coimbatore, Tamil Nadu, India

George A. Tsihrintzis
Department of Informatics
University of Piraeus
Piraeus, Greece

ISSN 2199-1073   ISSN 2199-1081   (electronic)
Internet of Things
ISBN 978-3-030-63939-6   ISBN 978-3-030-63937-2   (eBook)
https://doi.org/10.1007/978-3-030-63937-2

This Springer imprint is published by the registered company Springer Nature Switzerland AG
The registered company address is: Gewerbestrasse 11, 6330 Cham, Switzerland

Preface

With the advancement of Internet technology, the Internet of Things (IoT) has revo-
lutionized the complete automation setup in practical applications. Specifically, the
medical field is one of the chief beneficiaries of IoT, which is also represented as the
Internet of Medical Things (IoMT). The surge in the numerous types of diseases and
the lack of medical facilities, mostly in a rural scenario, have paved the way for
more research into the concepts of IoMT. The ultimate objective is to make health
facilities accessible to everyone irrespective of their geographical location. Remote
monitoring of the patients is one of the significant aspects of IoMT. Apart from
remote monitoring, several issues in the current medical setup can also be solved
with the help of IoMT technology. In spite of the numerous advantages, there is still
scope for improvement when it comes to the practical implementation scenario.
This is one of the significant motivations behind the origin of this book. The book
provides innovative ideas/research concepts which can enhance the practical feasibility of the proposed subject. Different practical applications are covered in this
book which will create an interest among engineers in these areas. This book,
indeed, will help readers to grasp the extensive point of view and the essence of the
recent advances in this field. A brief introduction about each chapter is as follows.

Chapter 1 is an introduction on IoMT with special emphasis on the various security aspects of this technology. The threats and challenges associated with the stor-
age/transfer of medical data are dealt with in detail. Few solutions for solving these
problems are also given, which will be an encouraging factor for anyone dealing
with IoMT. Chapter 2 deals with the overall concept of remote patient monitoring.
The different setups and automated systems necessary for remote patient monitoring are covered in this chapter. Global Positioning System (GPS) based patient
tracking system and Graphical User Interface (GUI) based medical data storage/
management/transfer are the highlights of this chapter. Chapter 3 deals with storage
and maintenance of Electronic Healthcare Records (EHR). Specifically, the security
issues in managing these confidential medical records are discussed in detail. The
inclusion of blockchain technology for solving this problem is the added advantage
of this chapter.

Chapter 4 covers the broad area of telemedicine. In the case of big data, such as medical records, there is always a problem in the transfer rate of data from the transmitter to receiver. Hence, there is a huge scope for compression methods in these telemedicine applications. Several lossless compression methods suitable for medical data are discussed in this chapter. Chapter 5 deals with the concepts of wearable technology. Wearable smart devices have become part and parcel of IoMT. Sensing of medical data from the human body is the major role of these smart devices apart from communication aspects. This chapter is specially focussed on wearable devices for cardiac disease detection and the related processing of data. Chapter 6 deals with the concept of assistive devices. IoMT-based assistive device setup for the elderly in homes is discussed in this chapter. Challenges faced by the elderly during emergency situations is the prime focus of this chapter.

Chapter 7 also deals with smart devices, but the application is different from the previous chapter. The focus of this chapter is on monitoring the vitals of the human body, such as blood pressure and temperature in a remote manner. Chapter 8 focuses on the remote monitoring system for patients with Parkinson's disease. This chapter gains significant attention since this disease affects the quality of life of patients drastically. Chapter 9 deals with diabetes, which is another common disease in the current scenario. Remote monitoring and detecting the intake of insulin using machine learning techniques are the focal point of this chapter.

Chapter 10 deals with the specific security vulnerabilities faced by any IoMT system. This chapter is a bit oriented towards the software aspects of any IoMT system, which is also equally important for practical applications. Chapter 11 deals with the concept of deep learning to tackle the security issues of IoMT systems. The integration of deep learning and IoMT is really worth exploring. Chapter 12 deals with the mathematical modelling of an IoT-based health monitoring system. This chapter is very important since IoMT is still under research and mathematical modelling is ideally suitable for the research scenario. Chapter 13 deals with the future challenges of IoMT-based setup for practical applications. Many ideas can be derived from this chapter which will be useful to develop innovative solutions.

We are grateful to the authors and reviewers for their excellent contributions in making this book possible. Our special thanks go to Prof. Fortino Giancarlo and Liotto Antonio (Series Editors of Internet of Things) for the opportunity to organize this edited volume. We are grateful to Springer, especially to Mr. Michael McCabe (Senior Editor), for the excellent collaboration. Finally, we would like to thank Ms. Olivia Ramya Chitranjan, who coordinated the entire proceedings. This edited book covers the fundamental concepts and application areas in detail, which is one of its main advantages. Being an interdisciplinary book, we hope it will be useful to a wide variety of readers and will provide useful information to professors, researchers, and students

Coimbatore, Tamil Nadu, India   D. Jude Hemanth
                                                                             J. Anitha
Piraeus, Greece   George A. Tsihrintzis
October 2020

Contents

**5   Wearable Smart Devices for Remote Healthcare Monitoring
to Detect Cardiac Diseases** 75

Ashok Kumar Munnangi, Ramesh Sekaran, Geetha Velliyangiri,
Manikandan Ramachandran, and Ambeshwar Kumar

Chapter 1
Internet of Medical Things: Security Threats, Security Challenges, and Potential Solutions

Amsaveni Avinashiappan and Bharathi Mayilsamy

1.1 Introduction

Due to technology advancement in machine learning, deep learning, Internet of Things, big data, and cloud computing, more powerful and advanced applications are blooming. Among a few applications, the Internet of Medical Things environment involving clinical sensors, clinical frameworks, and computing frameworks has drawn huge consideration prompting a few noteworthy upgrades in healthcare services. In an IoMT framework, the clinical sensors gather physiological information of patients and send them to the database store that can be acquired by approved healthcare experts (e.g., specialists, medical attendants, doctors) as shown in Fig. 1.1. The IoMT has changed the strategic vision of healthcare institutions as it has the potential to unlock endless gaps in the diagnosing, treatment, and maintenance of a patient's well-being and welfare, which holds the secret to reducing prices while enhancing clinical efficiency [1–5].

The IoMT may help in monitoring health of a patient on real-time basis and also assess the effect of treatment. IoMT can be a considered as a promising arrangement without clinical assets and help in avoiding unnecessary hospitalizations. About 60% of worldwide healthcare agencies have now adopted the Internet of Things technology, and a further 27% are projected to do so by 2021. IoMT will lessen clinical costs and elevate greater adaptability contrasted with conventional healthcare services [6]. The recurrence of contact among patients and doctors/nurses can be decreased altogether utilizing IoMT, particularly in the midst of pandemic (i.e., COVID-19).

A. Avinashiappan (✉) · B. Mayilsamy
Department of Electronics and Communication Engineering,
Kumaraguru College of Technology, Coimbatore, India
e-mail: amsaveni.a.ece@kct.ac.in; bharathi.m.ece@kct.ac.in

© The Author(s), under exclusive license to Springer
Nature Switzerland AG 2021
D. J. Hemanth et al. (eds.), *Internet of Medical Things*, Internet of Things,
https://doi.org/10.1007/978-3-030-63937-2_1

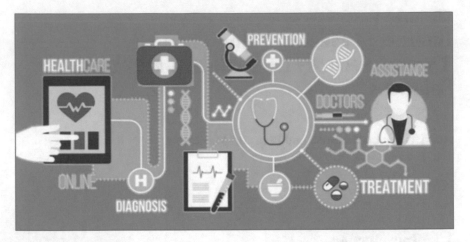

Fig. 1.1 Elements of IoMT

Although IoMT ensures reliable and effective healthcare services, it comes along with serious limitations like data leak. Specifically, a patient's clinical data gathered by clinical sensors is sent to data servers through open channel associated with the Internet. During the transmission, vindictive clients may wiretap (or listen in) the personal information, accordingly prompting data leak. Conventional cryptographic techniques like access control, identity authentication, and data encryption are usually employed to protect security during transmission through wireless communication. However, the computational resource requirements are not feasible due to limited memory space, computational capability, and energy capacity in case of medical sensors [7, 8].

1.2 IoMT Devices

More human healthcare institutions are utilizing the Internet of Medical Things to treat patients effectively and precisely. In the course of recent years, IoMT has seen noteworthy development. The Frost and Sullivan appraises that there were 4.5 billion IoMT gadgets in 2015, a number that they hope to increment to somewhere in the range of 20 and 30 billion by the year 2020.

Unfortunately, the growth in IoMT devices has resulted in an increase in vulnerabilities and open doors for these IoMT gadgets to get hacked. At the point when these smart gadgets aren't made sure about, it can put patients in danger and harm a human services association's whole framework.

IoMT would decrease needless doctor consultation and the pressure on healthcare by connecting patients with doctors and sharing of medical data across a safe network. The global IoMT market stood at $25 billion in 2016, according to Frost and Sullivan's analysis. This is expected to hit $75 billion by 2021, at an average annual rate of 30%. The range of IoT devices are growing each and each day. The reason for growing the wide variety of IoT devices is that they offer consolation in human life and perform work with higher outcomes than human beings. It's been said that, in 2018, the range of IoT devices may have greater than tripled considering 2012 and there can be 50 billion gadgets connected to the Internet. Figure 1.2 depicts the variety of connected IoT gadgets from 2012 to 2020.

The IoMT devices include wearable gadgets and vital monitors, specifically for use on the body, at home or in the community, in clinic or hospital environments, and in related real-time environments and virtual health and other services. These devices are summarized in Table 1.1.

1.2.1   On-Body Devices

The on-body devices may be widely classified into public health wearables and surgical wearables. Public health wearables include professional brands for personal exercise, such as fitness apps, athletic watches, smart apparel, and wristbands. Some of these items are non-approved by health agencies but may be authorized based on unbiased scientific tests and business research by professionals for limited safety applications. Samsung Health, Fitbit, Misfit, and Withings are businesses that work in this area [9].

Wearables devices include ECG monitoring patches, Biopatch to track a patient's condition minute by minute, insulin pumps, blood monitors, and wearables where doctors and nurses monitor for patient vitals and conditions. Wearables have turned

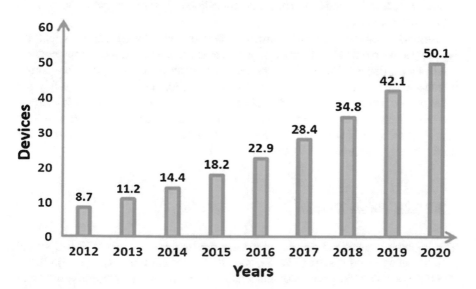

Fig. 1.2 Range of IoT devices from 2012 to 2020

Table 1.1 IoMT devices

On-body devices
Public health wearables
Surgical wearables
In-home device
Private emergency management system (PEMS)
Virtual patient management system (VPMS)
Virtual telehealth services (VTS)
Community devices
Mobility systems
Emergency management
Medical care devices
Kiosks
Logistics
In-clinic devices
Function devices
Therapeutic application devices
In-hospital devices
Product management
Emergency management
Resource management
Environmental and energy monitoring

into a "must-have" in the healthcare industry becoming more common, and usage is expected to increase.

Surgical wearables include approved products and network supports that are typically authorized to be utilized by health agencies. Any of similar instruments are used in accordance with specialist guidance or prescription by a physician. Types include Active Protective's smart belt that tracks slips and sets up hip safety for older people; Halo Neuroscience's Halo Sport helmet that is used through exercise and athletic activity to activate brain regions accountable for muscle control, agility, and stamina; and NeuroMetrix's Quell, a portable neuromodulation system that hooks through tactile nerves to alleviate chronic pain.

1.2.2   In-Home Devices

The in-home devices are private emergency management system (PEMS), virtual patient management system (VPMS), and virtual telehealth services (VTS). PEMS combines a wearable gadget/communication network and a live emergency call center to improve self-reliance for elderly who are homebound or have limited mobility. This kit helps users to connect easily and seek emergency medical services.

VPMS includes both patient monitoring systems and monitors utilized to treat chronic conditions, which include constant tracking of medical criteria to facilitate lifelong treatment in a patient's home in an attempt to delay the development of the disease; remote surveillance and continual monitoring of discharged patients to increase recovery period and avoid rehospitalization; and drug administration, to provide patients with prescription details and dosage reminders. VTS encompass online appointments that assist patients to handle their conditions and get medications or prescribed treatment plans. Examples involve video consultation and symptom or lesion assessment through video examination and virtual testing.

1.2.3   Community Devices

There are five types of community devices:

- Mobility systems allow passenger cars to track physiological conditions while in transit.
- Emergency management technology is intended to support medical professionals in the first aid, ambulance, and emergency room units.
- Kiosks are objects capable of providing products or services such as communication to caretakers, often with a mobile touchscreen displays.
- Medical care devices are equipment that are used outside the home by a manufacturer or conventional healthcare facilities, for example, in a healthcare camp.
- Logistics means mass transit and distribution of medicines and facilities, such as pharmaceuticals, medical and surgical materials, medical instruments and appliances, and other items required by healthcare professionals. Examples of IoMT include monitors in prescription packages that monitor temperature, pressure, humidity, and tilting; end-to-end visibility systems that control customized medication for a particular cancer patient utilizing radiofrequency identification (RFID) and bar codes; and drones to provide quicker last-mile distribution.

1.2.4   In-Clinic Devices

The in-clinic devices are tools which can be used for functional or therapeutic uses. There is one main difference between medical care devices and community, i.e., instead of manually accessing the system, the operator may be centrally controlled when the product is being operated by trained staff. Examples include the Rijuven's Bag Clinic, a cloud-based assessment tool allowing doctors to examine patients at any time; the Thinklabs automated stethoscope; and the Thyrocare Advanced Virtual

Medical Monitoring System for the head, lungs, mouth, neck, throat, and abdomen. It can also be used to diagnose patients' temperature.

1.2.5   In-Hospital Devices

In-hospital devices are categorized into IoMT-compatible devices and a wider category of systems in many management areas:

- Product management tracks costly equipment and mobile assets, such as infusion pumps and wheelchairs.
- Personnel management monitors the performance and profitability of the employees. Patient flow management enhances hospital efficiency by reducing bottlenecks and improving patient service—such as tracking patient arrival times from an intensive care unit to the post-care room.
- Resource management facilitates the procurement, storing, and utilization of hospital equipment, consumables, pharmaceuticals, and medical instruments to minimize production costs and enhance employee performance.
- Environmental and energy monitoring (such as temperature and humidity) monitor the use of electricity and ensure optimal conditions in patient and storage areas.

1.3   IoMT System Architecture

Figure 1.3 demonstrates the architecture of the IoMT system being considered. The architecture comprises three layers, namely, data collection layer, data management layer, and medical server layer. Healthcare sensors in the data collection layer gather vital parameters of the patients and transmit the data collected to the data servers

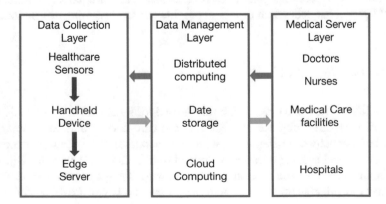

Fig. 1.3 IoMT system architecture

positioned in the data management layer following appropriate pre-processing on the edge servers that are installed in patient proximity.

The data servers in the data management layer acquire, process, and interpret the captured clinical data by supplying approved healthcare professionals (such as doctors) with access interfaces in the medical server layer. Upon collecting and reviewing patient clinical information on the medical server network, all healthcare professionals and clinicians provide patients with expert advice or take immediate action [10]. Here the three layers of IoMT are explained in detail.

1.3.1   Data Collection Layer

The data collection layer consists of healthcare sensors, handheld devices, and edge servers. Healthcare sensors are measurement systems which are implanted, worn, or attached to garments. They capture vital parameters of patients from various sources, including heart rate, temperature, respiration rate, blood glucose, blood pressure, oxygen level in blood, pulse rate, acceleration, and electrocardiogram (ECG).

The stringent observation should be carried out in emergency situations. Healthcare sensors will capture critical physiological data on an ongoing basis, which will be saved on a portable computer (such as iPhones, tablets, laptops). Whereas in typical situations, occasional observation will be conducted at the request of designated individuals (e.g., patient, doctor, or nurse). At regular intervals, the healthcare sensors can capture patient data. Using low-power wireless devices such as near-field communication (NFC), RFID, and Bluetooth Low Energy (BLE), each healthcare sensor is wirelessly connected to a specific mobile device. Upon pre-processing the gathered data, the mobile computer must send the data to the edge server which is typically installed near patients.

The edge servers are usually installed on wireless access systems (such as wireless routers, IoT gateways, and base stations) and compatibly communicate with various forms of local networks. The original heterogeneous data in various configurations and formats is pre-processed and analyzed internally at the edge server before sending it to the next layer. Edge servers are equipped with greater computing capacities and ample battery backup compared to healthcare sensors and mobile devices. Thus, edge servers may have better encryption algorithms to protect data transfer between the data collection layer and the data management layer. Furthermore, proposals for early detection and recovery progress evaluation may be done on the edge server with patient's permission. The edge servers will also produce warning signals when some critical physiological parameter exceeds a threshold, to warn the patient or doctor.

Owing to computing capacity constraints and the power supply of healthcare sensor systems and mobile systems, the security and privacy in the data collection layer face problems. Consequently, this layer's security measures must be less computationally intensive and generate less communication overhead.

1.3.2   Data Management Layer

The key purpose of data management layer is to handle the heterogeneous data obtained from the data collection layer in a standardized manner. The medical data obtained will be collected and evaluated based on the timeliness of the healthcare data and the importance of the tasks of the study. Data services and processing facilities are given inside this layer to save, process, and interpret the vital parameters gathered.

These centralized systems are usually combined with large data, cloud infrastructure, and storage technology for effective data collection, research, and storage. Accordingly, cloud-based data collection would dramatically minimize processing time, while cloud data management would offer the omnipresent ease of viewing data anytime and wherever.

Confidentiality of patient data is a critical concern, as long-term clinical records of patients are maintained on the computer servers. The data must therefore be secured when it is held at the data management layer. Complicated and complex data protection techniques can be implemented to maintain data confidentiality and patient privacy relying on the high computing capacity of data storage facilities (such as clouds). This layer also makes certain identification verification and access management frameworks feasible.

1.3.3   Medical Server Layer

Within this layer, healthcare practitioners (e.g., physicians, nurses) who have credentials for authorization and authentication can access patient physiological data. Following the process and research on patient medical data in data management layer, visual data descriptive findings in this layer are presented to healthcare professionals. If the patient's medical data changes, healthcare professionals can detect quickly and take corrective steps.

Clinical guidance from clinics and medical testing facilities to patients will reduce the amount of clinic visits and examinations. In fact, a patient's rehabilitation treatment can be tracked back to healthcare practitioners. This is useful for optimizing treatment. The primary concern to ensure the protection of patient data on the medical server layer is that only certified healthcare practitioners shall read the data. Access protection frameworks (like digital signatures or certificates) can therefore be developed to ensure that only intended users with access privileges may use the data. Medical data sharing guidelines to avoid leaking information about patients' privacy are also implemented in this layer. Table 1.2 summarizes the various security requirements and threats on IoMT systems.

Table 1.2 IoMT security requirements and threats

Layer	Security requirement	Security threat
Data collection layer	Access control, threat hunting, authentication	Insufficient physical security, inadequate harvesting of energy
Data management layer	Traffic monitoring, encrypt the payload, anomaly detection, traffic shaping	Improper encryption, unnecessary open ports
Medical server layer	Secure API, application verification, information forensics	Weak programming practices, insufficient audit mechanism, improper patch management

1.4   IoMT Attack Types

How do IoMT devices get hacked? Recently, hackers are more creative in their methodologies and use various tactics to break into a system. It is inevitable to consider the following threat vectors while designing an IoMT defense scheme [11–14].

Side Channel   This attack takes advantage of data leakage and has proven to be one of the effective and powerful attacks. Attackers can monitor the electromagnetic activity around the particular medical device and steal data.

Tag Cloning   This attack enables the attacker to duplicate or clone the data collected via side channel attack. It is easy for attackers to clone RFIDs.

Tampering Devices   This attack happens when the sensors are physically tampered. The functionality of the device may be partially or entirely stopped or manipulated. The device can be exploited through firmware vulnerabilities by installing malware in them.

Sensor Tracking   Most of the monitoring devices have GPS sensors that use a patient's physical location to facilitate emergency situations. There is high chance for attackers to hack into such devices to access patient's location and even can send inaccurate information.

Eavesdropping   In an attempt to locate a smart device, an attacker can intercept wireless communications through hardware devices. For instance, a patient's vitals may be intercepted on transmission, where the data can then be misused.

Replay   An authentic message exchanged previously between legitimate users can be rescued by attackers. OneTouch Ping insulin pump was detected with this vulnerability in 2016 due to the absence of secure communication mechanisms.

Man-in-the-middle   This attack is when an attacker intrudes or alters data shared between two parties involved in the transmission. An altered data leads to miscommunication and in turn may provide false results leading to mistreatment.

Rogue Access This attack is when an attacker creates a forged gateway in the wireless network to grant access to unauthorized users. The attacker can intercept the traffic without being identified.

Denial-of-service (DoS) This attack is when an IoMT device is flooded with service requests. This will disrupt the availability. IoMT devices in botnets can be hacked and to use infected devices without being identified by the owner.

Cross-Site Request Forgery (CSRF) An attacker can trick the end user into acting on an insecure vulnerable application without being identified by the user. The IoMT device layer's web interface becomes vulnerable to this attack, and hence it must be configured properly.

Session Hijacking For smart gadgets that handle session association at the web interface level, the interfaces become powerless against session hijacking. This permits an assailant to assume control over the session particulars and control it.

Cross-site Scripting (XSS) This exploits IoMT by infusing uniquely made contents to sidestep access controls through website pages. The web interface of IoT gadgets associated with the cloud is defenseless against these assaults.

SQL Injection A SQL injection assault is where an aggressor executes noxious SQL statements to sidestep the gadgets security efforts where it can compromise delicate patient information or modify basic information.

Account Hijacking An aggressor can perform account hacking by capturing the transmission performed between IoMT parts while an end client is being verified. The ascent of these assaults is due to the effect of gadgets that have unpatched weaknesses.

Ransomware During a ransomware assault, programmers scramble information like patient records and hold it in return for cash. This danger can begin with only one machine and then, at some point, spread all through the system. This attack could likewise be effective by denying access to IoMT gadget productions, compromising patient security until money is paid.

Brute Force This is the most effortless feasible way assailants can access a data server and IoMT gadgets since little assurance is set up to prevent such assaults in IoT gadgets.

1.5   Challenges in IoMT Security Schemes

All layers of a three-layer IoMT may exhibit security challenges, while the vulnerabilities in data collection layer are severe compared to the other layers. The root cause for vulnerabilities can be the limitations in memory space, computation capability, and energy supply. One can confront additional trouble in structuring feasible security plans as the clinical sensors exhibit variety and mobility characteristics [15, 16].

It is important to consider small form factors like size and volume of medical sensors. The portable size adds to the convenience in user applications and may be less expensive, but will include the limitations of memory space and power efficiency. The memory space may not be enough to provide space for complicated security protocols performing complex operations leading to delay in communications, which can be dangerous for patients. Hence it is not advisable to rely on current security algorithms for normal operations with limited resources. The security plot appropriate for this gadget ought to involve as less resources as possible, and the necessities like storage space and computing capability of the clinical sensors ought to be sufficiently low without influencing sensor operations.

Although clinical sensors accompany explicit preparing limit, they choose to use less power because of confined battery limit. Without fundamental information to process, they will work at a lower CPU speed to save power use. When it isn't essential to report sensor readings, such devices will operate in power sparing mode. Furthermore, for certain chips (i.e., ICs) embedded into the human body, swapping them for the absence of power intensity will realize both distress and critical cost. Along these lines, the energy requirement of clinical sensor prompts the difficulties in designing security plans.

There are different sorts of clinical sensors, from developed PC gadgets to RFID labels and even chips implanted in the human body. The processing capacity, capacity, memory, and inserted programming of every kind of sensor are unique. In this way, structuring a security that can oblige even the gadgets with most vulnerable ability is another challenge.

Generally, clinical sensors are versatile rather than being static. The versatility or the mobility of clinical sensors enhances the relevance of IoMT. For instance, patients can direct physical exercises in limited clinical region while being persistently checked. Moreover, mobility empowers patients to move from their family rooms to different spaces for clinical assessment without intruding on the persistent clinical monitoring. Likewise, clinical sensors may associate or leave the IoMT network at any second, in this way inciting the dynamic changes of the framework topology joined by challenges in the focal administration or secret key conveyance. In this manner, it is additionally a genuine challenge to plan a safety plot that fulfills the portability prerequisite.

The occurrence and severity of threats on healthcare organizations—as well as the risk to confidentiality of patients and safety—means that companies come under pressure when it comes to medical instrument security. Because of recent incidents, including the WannaCry ransomware threat, lawmakers highlighted the severity of cybersecurity problems that threaten legacy applications and the equipment. The FDA (Food and Administration) has already made guidelines for product manufacturers, but companies do not have to comply since these instructions are not legitimate requirements.

1.6   Current Security Plans for IoMT

Security concerns over IoMT can be hazardous as vital parameters of patients can be abused prompting social separation or maltreatment against patients. Authentication, data encryption, and identity authentication are approaches helping to preserve data security in IoMT devices [17–19].

Access control prevents illegal access to data by unauthorized sources. It gives appropriate degree of consent to approved sources. It gives access in two modes in particular quality attribute-based access and glass-breaking access. The attribute-based access is required for fine-grained control if the circumstance is ordinary. The glass-breaking access can be utilized in crisis circumstance. In this method, patients themselves code the gathered clinical information. IoMT information storage framework is planned with twofold adaptable access control component. The component is self-versatile under ordinary situation and further ensures medical aid in crisis situation. In attribute-based encryption calculation in IoMT, the privacy concerns have been addressed through encryption, which in other terms mean hiding of the addressed data. The access control scheme in this kind of framework will help in preventing leaking of data from the data storage servers.

Validating identity can prevent unauthorized users from accessing sensitive data, and this method is called identity authentication. It verified the integrity between the patient and confidential physiology data. This will improve information security through validation and key agreement protocol that gave an access control component. Also, a handshake procedure and session resumption strategy-based start to finish plot is intended for protecting the IoMT framework. Thus, the IoMT framework is based on the authorized and authenticated architecture design considering the portability boundary of clinical sensors. The authentication instrument in the framework has anonymous certificateless aggregate signature scheme. An anonymous attribute-based encryption will enable the access control mechanism. In two-way identity authentication method, both the keys, i.e., authentication and session keys, will provide identity verification. The legality of heterogeneous medical sensors can be authenticated through this two-way method. The two mechanisms, elliptic curve encryption algorithm signature and symmetric encryption algorithm of session key, will help in achieving identity authentication. The anonymous authentication plan has been developed to help in preventing the authenticated patients from untrusted authentication.

Data encryption is suggested for the storage servers in management layer of IoMT architecture. Data encryption is a strategy which empowers the two players to transfer information based on concurred rules. It helped to realize secured data transmission through Advanced Encryption Standard (AES)-based key distribution scheme in IoMT. Additionally, matrix scheme based on homomorphic encryption that will enable privacy-preserving strategy is also used. The security of data can be ensured at the time of data collection. To enhance security during acquisition and transmission, two algorithms, namely, lightweight FPGA hardware-based cipher algorithm and secret cipher share algorithm, can be used. It guaranteed integrity and confidentiality of data transmission through D2D-assist data transmission protocol. Certificateless generalized signcryption technique is used in this protocol [20, 21].

Most of the currently used security plans are efficient in data management layer and medical server layer in case of three-tier architectures. It empowers productive processing facilities that can bolster complex computations and energy-consuming errands like decryption, encryption, data analytics, and more. Whatsoever the security in data collection layer is vulnerable compared to data manage layer and medical server layer. As the quantity of clinical data generated is massive, using lightweight encryption methods also will be expensive due to resource strained clinical sensors. The encryption methods introduce communication delays that may be considered a potential threat to patients in emergency scenario. Hence access control, authentication, and data encryption mechanisms are not feasible because of their constrained computational ability, memory space, and power flexibility.

1.7 Potential Solutions for Security Vulnerabilities

As there is an assortment of ways, hackers can discover their way into the clinical devices and contrarily sway clinical tasks, which is the reason guaranteeing that these gadgets are secure from the present dangers is pivotal for each social healthcare provider [22]. The following are the solutions used to safeguard against these assaults:

- **Security experts available 24×7×365**: IoMT gadgets and attackers are working nonstop, where attackers are reliably finding better approaches to steal data. Having security specialists checking the system all time, permits in-house security group to concentrate on other IT anticipates, and also can alarm when there is a threat entering the system which help in remediating the danger.
- **SIEM**: A security information event management (SIEM) solution can give a healthcare provider more noteworthy perceivability into their systems and comprehend their system foundation and the distinctive IoMT gadgets associated with it, making it simpler to distinguish dangers before gadgets can be undermined.
- **Patch management**: Not all IoMT gadgets can be fixed for an assortment of reasons. For those that can, fixing your IoMT gadgets can kill the framework's weaknesses. At the point when you implement the procedure, make it a point to

note weak gadgets that can't be fixed and that the remaining IoMT gadgets are appropriately fixed to guarantee safety and avoid latest threats.

- **Devices with default passwords**: Set unique, strong credentials for all devices and services.
- **Ethical software**: Review gadgets for rogue programming and perform uninstallations as suitable; limit authorizations to forestall future rogue installations.
- **Authorized network access**: Design the network access control framework with better characterized and more vigilant security approaches.
- **Avoid device misuse**: Limit web browsing to pre-affirmed whitelisted destinations, permitting new destinations upon demand.
- **Malicious activity**: Progressing surveillance of your IoMT system to proactively distinguish and fix possible vulnerabilities, decreasing the probability that assailants can compromise the framework.
- **Lack of containment**: It's critical to not just get ready to repulse assaults before they land, however, to have controls set up that permit you to contain and thwart them should they go through your barriers. Until this end, you should build and implement a segmented system not only at the edge, yet inside around endpoint bunches that share comparative clinical applications and system work processes.

1.8   Conclusion

Conventional health systems are undergoing a massive shift because digital technology brings high-tech and interactive devices in the hands of customers and provides patients and doctors with greater access to health resources, including in the lowest and most isolated regions. When the usage of IoMT systems in most healthcare organizations and personal lives grows, security issues are growing. A broad variety of threats have arisen due to resource shortages and device complexity in diverse IoMT environments. Many of these threats will contribute to device failure in the IoMT environment. As there is no predefined model of IoMT environment, IoMT security problems and approaches without a clear framework have been raised by much of the researchers. This chapter offers an extensive review of IoMT security specifications, problems, and approaches focused on the security features of the popular three-layer structured IoMT environment. The key purpose of this review was to provide the IoMT professionals and analysts a taxonomy of IoMT cyber challenges to carry out their potential work focused on growing problem and response.

References

1. Abomhara, M., & Koien, G. M. (2014). Security and privacy in the internet of things: Current status and open issues. *IEEE International Conference on Privacy and Security in Mobile Systems, 2014*, 1–8.
2. Chen, S., Xu, D., Liu, B., & Wang, H. (2014). A vision of IoT: Applications, challenges, and opportunities with china perspective. *IEEE Internet of Things Journal, 1*(4), 349–359.
3. Rafique, W., Qi, L., Yaqoob, I., Imran, M., Rasool, R. U., & Dou, W. (2020). Complementing IoT services through software defined networking and edge computing: A comprehensive survey. *IEEE Communications Surveys Tutorials, 22*, 1–45.
4. Bharathi, K. S., & Venkateswari, R. (2019). Security challenges and solutions for wireless body area networks. In *Computing, communication, and signal processing* (pp. 275–283). Singapore: Springer.
5. Sangaiah, A. K., Dhanaraj, J. S. A., Mohandas, P., & Castiglione, A. (2020). Cognitive IoT system with intelligence techniques in sustainable computing environment. *Computer Communications, 154*, 347.
6. Abawajy, J. H., & Hassan, M. M. (2017). Federated internet of things and cloud computing pervasive patient health monitoring system. *IEEE Communications Magazine, 55*(1), 48–53.
7. Xiong, N., Vasilakos, A. V., Yang, L. T., Song, L., Pan, Y., & Kannan, R. (2009). Comparative analysis of quality of service and memory usage for adaptive failure detectors in healthcare systems. *IEEE Journal on Selected Areas in Communications, 27*(4), 495–509.
8. Syed, L., Jabeen, S., Manimala, S., & Alsaeedi, A. (2019). Smart healthcare framework for ambient assisted living using IOMT and big data analytics techniques. *Future Generation Computer Systems, 101*, 136–151.
9. Yaacoub, J. P. A., Noura, H. N., Noura, O., Salman, E., Yaacoub, R., Couturier, & Chehab, A. (2020). Securing internet of medical things systems: Limitations, issues and recommendations. *Future Generation Computer Systems, 105*, 581–606.
10. Xuran, L., Hong-Ning, D., Qubeijian, W., Muhammad, I., Dengwang, L., & Muhammad, A. I. (2020). Securing Internet of Medical Things with Friendly-jamming schemes. *Computer Communications, 160*, 431.
11. Somasundaram, R., & Thirugnanam, M. (2020). Review of security challenges in healthcare internet of things. *Wireless Networks, 7*, 1–7.
12. Sandeep, P., Oluwarotimi, S., Wanqing, W., Arun, K., & Guanglin, L. (2019). A joint resource-aware and medical data security framework for wearable healthcare systems. *Future Generation Computer Systems, 95*, 382.
13. Mohan, A. (2014). Cyber security for personal medical devices internet of things. *IEEE International Conference on Distributed Computing in Sensor Systems, 2014*, 372–374.
14. Yeh, K. H. (2016). A secure IoT-based healthcare system with body sensor networks. *IEEE Access, 4*, 10288–10299.
15. Zhang, Y. M., Qiu, C., Tsai, M., Hassan, M., & Alamri, A. (2017). Health-cps: Healthcare cyber-physical system assisted by cloud and big data. *IEEE Systems Journal, 11*(1), 88–95.
16. Gatouillat, A., Badr, Y., Massot, B., & Sejdić, E. (2018). Internet of medical things: A review of recent contributions dealing with cyber-physical systems in medicine. *IEEE Internet of Things Journal, 5*(5), 3810–3822.
17. Cao, R., Tang, Z., Liu, C., & Veeravalli, B. (2020). A scalable multicloud storage architecture for cloud-supported medical internet of things. *IEEE Internet of Things Journal, 7*(3), 1641–1654.
18. Qureshi, F., & Krishnan, S. (2018). Wearable hardware design for the internet of medical things (IOMT). *Sensors, 18*(11), 3812–3820.
19. Chen, M., Ma, Y., Li, Y., Wu, D., Zhang, Y., & Youn, C. (2017). Wearable 2.0: Enabling human-cloud integration in next generation healthcare systems. *IEEE Communications Magazine, 55*(1), 54–61.

20. Mukhopadhyay, S. (2015). Wearable sensors for human activity monitoring: A review. *IEEE Sensors Journal, 15*(3), 1321–1330.
21. Sun, Y., Lo, F., & Lo, B. (2019). Security and privacy for the internet of medical things enabled healthcare systems: A survey. *IEEE Access, 7*, 183339–183355.
22. Zhang, Y., Zheng, D., & Deng, R. (2018). Security and privacy in smart health: Efficient policy-hiding attribute-based access control. *IEEE Internet of Things Journal, 5*(3), 2130–2145.

Chapter 2
Intelligent Transit Healthcare Schema Using Internet of Medical Things (IoMT) Technology for Remote Patient Monitoring

R. J. S. Jeba Kumar, J. Roopa Jayasingh, and Deepika Blessy Telagathoti

2.1 Introduction

The introduction of emergency vehicles dates from the fourteenth century, when wagons were used to transport patients to the medical clinic. Mortality rates associated with motor vehicle collisions and other traffic accidents are affected by the response time of emergency vehicles and the transport time of patients to hospitals for life-saving treatment [1]. Such delays should be reduced to improve the outcomes of accident victims. This can be achieved by implementing smart health care systems using the Internet of Medical Things (IoMT), as described in this chapter.

A report by Fortune Business Insights (19 February 2020) highlighted the significant growth drivers, opportunities, restraints, key industrial insights, and competitive platform of IoMT. IoMT has a number of benefits, including improved drug management, real-time monitoring, efficient patient recovery, and reductions in medical expenses. Additional advantages are related to smart medical devices coupled with the increasing research and improvements to advanced technological networks. This additional awareness is expected to drive high profits during the near future.

Sensor-oriented devices that are integrated with mobile technology comprise IoMT. This technology has the potential to include patients, hospital staff, and visitors in the concept of smart healthcare and hospitals. The field of IoMT is very large-scale industry and has enormous benefits in relation to the development of the technology. A major advantage of the IoMT industry is improved patient assistance, better outcomes in medical treatment, enhanced manageability of medical

R. J. S. Jeba Kumar · J. Roopa Jayasingh (✉) · D. B. Telagathoti
Department of ECE, Karunya Institute of Technology and Sciences,
Coimbatore, Tamil Nadu, India
e-mail: roopa@karunya.edu

© The Author(s), under exclusive license to Springer
Nature Switzerland AG 2021
D. J. Hemanth et al. (eds.), *Internet of Medical Things*, Internet of Things,
https://doi.org/10.1007/978-3-030-63937-2_2

treatment, improved medication adherence, reduced medical errors, and decreased medical expenditures.

Future potential improvements to healthcare include rapid and efficient patient care, improved preventive care, better patient engagement and satisfaction, better care provision, and cost effectiveness for a mass implementation. IoMT has a great effect on our daily lives. Instead of visiting a hospital, a patient's health-related information can be remotely monitored and evaluated in a real-time data medical services system and then transmitted to a third party for further use via the cloud. An IoMT network is an extremely information-intensive domain with a perpetually increasing rate that requires safeguarding of an enormous amount of data without interference. Block chain is basically a digital ledger signifies the ability of a block chain integration to help for block based storage of data chunks of health monitored data that facilitates peer-to-peer communication. However, there are few disadvantages of IoMT, including the implementation of software for medical analysis. Physicians need to be highly trained to use the apps and the technology efficiently. Security challenges also exist as wireless devices are integrated in the field of the network. These challenges can be overcome by strongly encrypting the devices and system security.

This chapter discusses existing conceptualizations and automations used in IoMT systems for ambulance service [2, 3]. Telemedicine helps to bridge the gap between the patient and physician. IoMT is basically a network that comprises interconnected medical devices and technology that allows the devices to communicate over a network. The devices can transmit data across a common network, such as wi-fi. Sensors in this devices can capture vital health data, thus improving healthcare delivery and consequently a patient's outcome.

The Global System for Mobile Communications (GSM) is a digital mobile networking platform that uses time division multiplexing access for transmitting small data, such as Short Messaging Service (SMS) in portable telephone schema. GSM is increasingly being used as high-tech cableless telephone for empowering mobile connections. Global Positioning System (GPS) uses satellites for localizing a mobile device activated with a GPS module. Satellites send the precise latitude and altitude of the GPS-empowered device using trilateration and dynamic triangulation techniques. Details on a patient's medical condition can be transmitted to a specialist over the administrator site using IoMT schema.

This chapter discusses methods to improve patient outcomes by enhancing the traditional rescue vehicle approach. For example, GPS-GSM can be used after an accident by rescue vehicles in transit to the clinic. This chapter examines emerging functionality that tracks accidents close to a hospital for ambulance services, which can be equipped with smart mobile medical devices for data on electrocardiography (ECG), temperature, heart rate, and respiratory rate.

2.2   Impact of Industry 4.0 on IoMT Technology

Industry 4.0 includes innovative enhancements to medical devices and equipment. It allows for a computerized emergency clinic and a total evaluation framework that satisfies the individual needs of the patient/clinic with time and cost savings. Industry 4.0 produces high-caliber clinical gadgets that can be modified according to specific prerequisites. This transformation embraces mechanization to provide new opportunities in the clinical world. With the assistance of the connected Internet of Things (IoT) and the Integrated Operating System (IoS), it creates another virtual world. It facilitates the exchange of information by new components, programming, sensors, robots, and other propelled data advancements.

IoMT is an environment of associated sensors, wearable gadgets, medical devices, and clinical frameworks. It allows different human services applications to decrease social insurance costs, provide appropriate clinical responses, and target medical treatment. Driven by progress in remote communication, sensor systems, mobile phones, big data analysis, and distributed computing, IoMT is changing the human services industry by providing customized medical treatment and standardizing the communication of clinical information. IoMT has great impact with the advent of Industry 4.0, suh as the integration of the automobile industry with IoMT. The digital pillars of Industry 4.0—augmented reality, virtual reality, artificial intelligence, machine learning, and sensors—are shaping the future of the medical world. The embedded sensors in the IoMT paradigm enable the continuous monitoring of critical parameters such as ECG, heart rate, blood sugar level, temperature, and respiration. The integration of medical embedded devices, equipment, and the internet allows for the implementation of smart intensive care units (ICUs), smart thermal scanners, medical asset administration, and smart pill dispensers.

Although these clinical gadgets offer some advantages, they likewise raise genuine security and privacy concerns. Medical services frameworks gather and process sensitive and frequently vital clinical data. Cybercriminals focusing on the weaknesses in these IoMT gadgets mere increase of IoMT sensor nodes leads to data security and hence care should be taken while adding the node i.e. during scalling up of the system to avoid data loss of private health monitored sensitive data. Attacks on these devices can cause great harm to the patients.

Lab-on-a-chip is a computerized, scaled-down research framework that can be used inside and outside of a clinic for a broad range of patient tests, such as blood gases, glucose, and cholesterol levels. This innovation allows for rapid diagnostics with only modest quantities of tests and materials required. Industry 4.0 smart manufacturing and recent trends include manufacturing of wearable smart fabrics, for example, advanced medical bandage, by implementation of cross-domain knowledge by integrating medical and engineering field. IoMT has the potential to change the specialist consultation to interact with the patient seamlessly. The execution of Industry 4.0 will surely be a boost to the medical device industry considering its consistency and demonstration of quality frameworks. However, this unique area requires cutting-edge security research in the field of IoMT. Researchers and industry should commit to investigating novel security and privacy arrangements that address the specific needs of medical services.

2.3   Research Motivation

The key catalyst for this innovative research originated from the statistics in the media about the exponential number of deaths related to traffic accidents on national highways. In India, approximately one death occurs every 20 minutes on a high-speed road, with 1214 road accidents occurring every day. The National Crime Records Bureau Ministry of Home Affiars reported an average fatality rate of 11.6% per 100,000 people in 2018. The estimated overall number of traffic collisions in India was approximately 467,000 in 2018. Unfortunately, 30% of related deaths occur due to delayed ambulance services. The Indian government reported that more than 50% of heart attack patients reach the hospital late due to various reasons, including the unavailability of ambulance services and dense traffic on the roads. These issues with rescue vehicles are not found only in India; similar problems were reported for western nations by the British Broadcasting Corporation.

This chapter examines an intelligent transit healthcare (ITH) scheme using GSM and GPS satellite innovations. Smart rescue vehicles can perform fundamental tests for the patient using personal health monitoring gadgets in advance of arrival at an emergency clinic, such as ECG and body temperature. This chapter also discusses the use of IoMT to transmit a patient's health data to the treating physician prior to arrival at the emergency clinic. In this way, ITH-IoMT can save precious time and enable early, remote monitoring of a patient.

2.4   Literature Survey

Pradeep Kumar et al. proposed an emergency vehicle framework in which rescue vehicle lag is identified using Android technology and ZigBee. However, the main disadvantage of this approach is that the system is limited to a restricted width of the inclusion run as there is no specified boundary condition and hence cannot be tested or implemented for real-life application [4]. Sultana et al. proposed a Zigbee-based ambulance schema focused on ambulance tracking, but they did not address the critical factor of patient monitoring [5]. Loubet et al. proposed a health monitoring schema, although it lacked a clustering approach for broader application in which the patient condition is transmitted to a monitoring station. The devices, sensors, and architecture were interconnected using a public approach such as a wireless sensor network, which lacks data security; furthermore, the harvested data were not used for real-time monitoring [6].

2.5   Gap Analysis of Existing Work

The existing methodology for implementing the IoMT is limited to a specific range of geographical coverage. It fails to capture ad-hoc patient health details and does not protect the patient's private data that as classified under the Personal Health Information Protection Act of 2004 due to the use of unsecure wireless sensor networks that are prone to data leaks [4–6]. In the existing literature, the implementation of ZigBee or low-radiofrequency devices limits the coverage or efficiency of the overall system for practical implementation in a wider area. The ad-hoc patient data are not available to physicians and treatment facilities because currently the medical devices are not connected to the cloud. The existing work focuses on integrating GSM-GPS to track accident zones and not the health devices inside an ambulance to track a patient's health status. The implementation of interconnected nodes over a geographical area in an unsecure manner makes the existing work susceptible to data leaks of private health information. The previously mentioned studies did not incorporate prime functionality; thus, there exists a gap for real-time implementation and hence the need for a robust ITH-IoMT system is proposed in this chapter.

2.6   Proposed Intelligent Transit Healthcare Schema Using IoMT Networking System

The proposed intelligent transit healthcare system drastically minimizes the rate of deaths due to traffic collisions, as it ensures that ambulance services are provided immediately. It also informs the receiving hospital automatically. The smart ambulance contains the necessary equipment fitted with strong internet connections to transmit the details to the physician's administration page using IoMT connectivity. Two modules exist in this smart ambulance scheme: the module embedded in the patient's vehicle with GSM and GPS-enabled gadgets and a mobile phone-based cloud integrated module. A short message of the accident location is sent to the emergency clinic, activating a rescue vehicle.

The fundamental modules required for this ITH schema are GSM-GPS for data on the accident location, a transmitter and receiver for sending and receiving SMS, vibration sensors in the vehicle for accident recognition, a microcontroller as a handling unit, and an interface for ITH-IoMT to communicate with the GSM collector module. These modules are required to allow the ambulance services to assist an accident victim or others traveling to a hospital for health issues. Another set of portable devices are installed in the smart ambulance, including a chest strap (Qardiocore manufactured by Qardio company) embedded with sensors to record the heart rate, ECG, respiratory rate, and body temperature and other point-of-care devices (POC) used for diagnosing and monitoring purposes. The overall schema architecture for an ITH emergency vehicular system (EVS) is shown in Fig. 2.1.

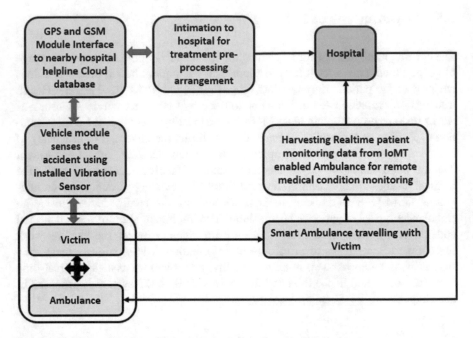

Fig. 2.1 Intelligent transit healthcare schema for intelligent transit healthcare

2.6.1 Vibration-Sensing Methodology for Accident Detection

A vibration-detecting framework is used to identify an accident event. When mechanical vibrations are detected, an electrical signal is amplified through an operational amplifier. A rectifier diode is used to limit the electrical flow, and this signal is moved over a circuit of comparator where another operational amplifier is used. The electrical signal travels to the bipolar semiconductor-based transistor that is used for exchanging intentions. Contrarily cut sign is moved over a timer for postponing reason. After this, a BC 547 semiconductor triggers a warning while the microcontroller lingerie its strategy for the pre-settled action dumped in microcontroller for mishap. The vibration-detecting framework is represented in Fig. 2.2.

2.6.2 System Safeguards

Revoke switch functionality to reset the system entirely if any false trigger due to system fault is being encountered. This procedure is used when the accident-related injury is not severe or to prevent wrong information because of framework breakdown to evade the plunder. Figure 2.3 shows the circuit arrangement for the revoke switch in which the Micro Controller Unit (MCU) unit triggers the whole schema.

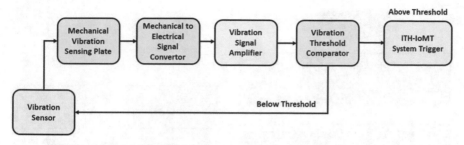

Fig. 2.2 Vibration-sensing framework for the ITH-IoMT system

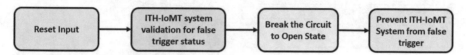

Fig. 2.3 Stability layout for ITH-IoMT to avoid system failure

2.6.3   GPS Integration

A GPS module is included in the automobile to capture the location data of scope and longitude from a situated satellite [7, 8]. The GPS module communicates with four satellites and detects the followed area with three-dimensional topographical zones. A flowchart depicting the GPS initialization is shown in Fig. 2.4.

2.6.4   Hospital Communication About Accident Location

Two gadgets are used in ITH-IoMT: a transmitter unit and receiver unit. The module in the EVS acts as the transmitter to send the latitude and longitude coordinates to the recipient. The receiver unit is portable and can be coupled by using a dual-communication integrated circuit to a personal system or a screen. This screen is used to activate the auto-dialer to direct the smart ambulance to the accident location by VB 6.0. Latitude and longitude coordinate positions along the streets and expressways can be seen on the personal system or screen to gather information communicated by the transmitter module from the accident scene. The emergency clinic or EVS is consequently dialed using the portable device on the collector side. This entire process occurs without any human intervention. Both the vehicle transmitter module and EVS module use Active Trade (AT) orders [9]. These AT orders are essentially used to build up the correspondence in type of little area detail to the close off medical clinic for human services administration.

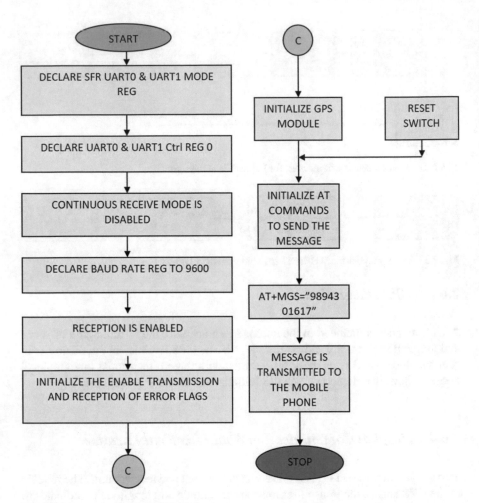

Fig. 2.4 Flowchart for the GPS initialization process

2.6.5 MCU Connection with the ITH-IoMT Subsystem

The GSM design is executed with the assistance of a mobile phone to provide data about the accident location. The GPS radio wire and GPS beneficiary are used for harnessing the medical scope, longitude coordinate, and altitude coordinate of the casualty's vehicle, which has been embedded in this model. An alert is used to dispatch the EVS. The MCU facilitates every one of these issues. This specific MCU has two sequential ports that are currently uncommon in other micro-regulators. A diagram of the ambulance scheme is shown in Fig. 2.5.

A smart ambulance has a GSM recipient module that is associated with the clinic database, interconnected with the emergency vehicle schema as shown in Fig. 2.6.

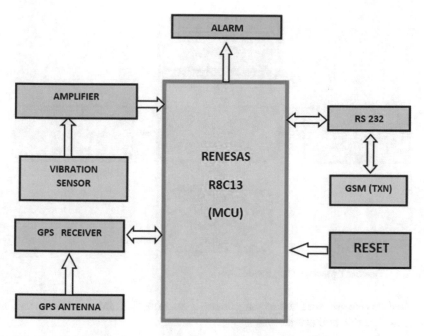

Fig. 2.5 Functional diagram of the ITH-IoMT ambulance scheme for the vehicle module

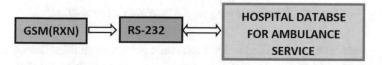

Fig. 2.6 Functional diagram of the ITH scheme connected to EVS

This ITH-IoMT framework is interconnected with the microcontroller unit for each sub-gadget, such as the vibration-sensing framework, GPS module, and GSM module as depicted in Fig. 2.7. An IoMT-equipped smart ambulance reduces the time delay between the ambulance's arrival and the victim's treatment [10–12].

2.7   Results and Discussion

The outcomes associated with ITH-IoMT mainly fall into two categories. First, the vibration-sensing system detects the occurrence of an accident and shares its location using a GPS-activated module in the victim's vehicle. Second, the smart ambulance reaches the accident scene, after which the victim's basic health data are securely transmitted to a nearby hospital using IoMT technology in the EVS. These data can be continuously monitored by the treating physician on the established

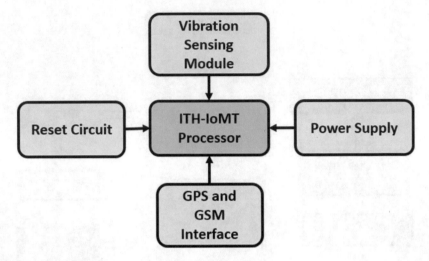

Fig. 2.7 ITH-IoMT processor integration layout

medical device network while the patient is in transit. Point-of-care devices are integrated in this ambulance system.

2.7.1 *GPS-Enabled Module for Location Information*

A GPS system can be installed in a vehicle, on a phone, or on other devices, which can either be fixed or mobile units. GPS provides data on a specific location and can also follow the movement of a vehicle or individual. A GPS framework uses the Global Navigation Satellite System (GNSS), which joins a scope of satellites that send microwave flags to GPS gadgets to provide data on the location, vehicle speed, time, and direction. The vibration sensors that are embedded in the victim's vehicle detect an accident and send the details to the nearest hospital. Figure 2.8 portrays the GPS Graphical User Interface (GUI) system portal. The accident location tracking template is shown in Fig. 2.9.

2.7.2 *ECG and Health Data Monitoring Schema*

An ECG screens the electrical flow through the heart to diagnose an assortment of cardiovascular issues, including arrhythmias, heart attacks, heat aggravation, heart failure, helpless blood gracefully, coronary corridor infection, and others. A heart injury (myocardial wound) is one of the most serious complications of an automobile accident. It occurs when an individual experiences blunt force trauma to the chest, which causes the heart muscle and myocardial tissue to become injured. Such

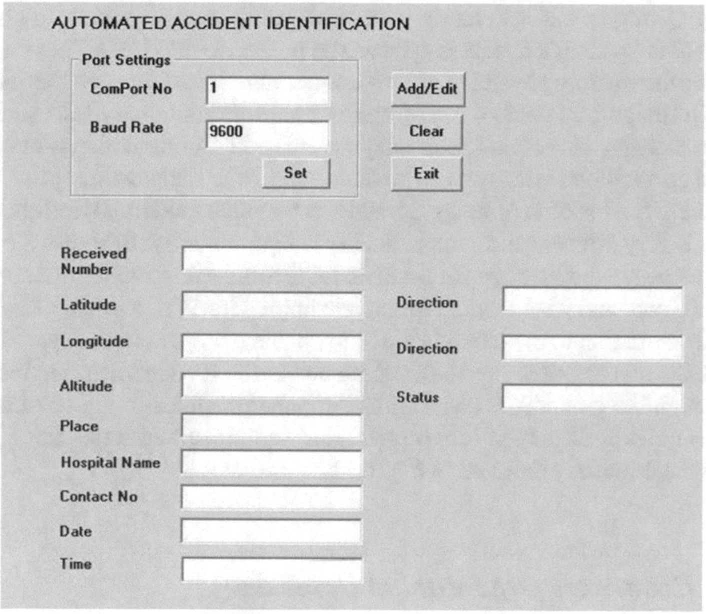

Fig. 2.8 Layout for the GPS-GUI system

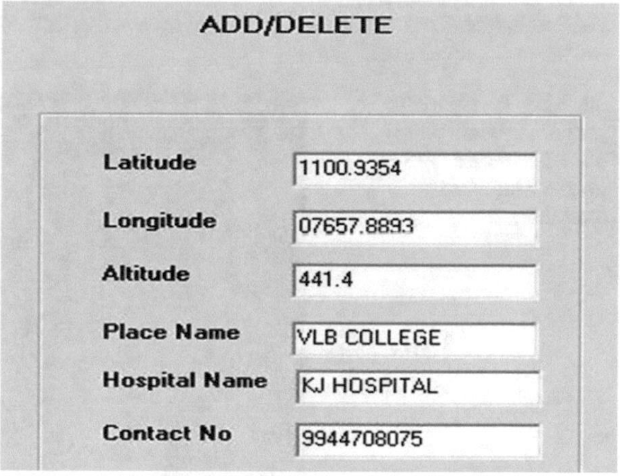

Fig. 2.9 Accident location captured via GPS-GUI

an injury may be caused by a progression of conditions during an accident; however it generally occurs when an individual was not wearing a safety belt in a high-speed collision, was injured by the force of an airbag, or was injured by the absence of an airbag. An ECG can determine the seriousness of heart injuries and direct treatment for these conditions.

Qardiocore is a portable device with embedded sensors designed to continuously monitor a patient's ECG, temperature, heart rate, respiratory rate, and activity tracking. Qardiocore is wirelessly connected using a wi-fi module. Internally, a database of the patient's details is created and pushed to the cloud. These details are viewed by the treating physician as they are communicated through the IoMT platform. POC diagnostic devices are primarily used for preliminary tests such as glucose, electrolytes, cholesterol, and enzyme analysis. Cardiomarkers, blood gases, and fecal occult blood tests also can be done using POC diagnostic appliances. POC devices have the benefits of being portable, convenient, and quick to process; connectivity is very efficient and sample quality is high. Figure 2.10 shows the comprehensive hardware architecture for portable ECG using a micro-controller unit, pulse monitoring pad, and wi-fi module to activate unified IoMT networking for transmitting the information to the cloud [13–15] to enable remote monitoring.

The ECG and essential health data obtained in the EVS are directed to the database through the cloud TCP convention. Any abnormalities in the collected information is quickly displayed in the physician's administrator page along with the patient's database, as portrayed in Fig. 2.11.

2.7.3 Healthcare Organization Database

Fundamental patient data such as heart rate, respiratory rate, and body temperature are pushed to the physician's user interface via IoMT. The entire ITH-IoMT framework enhances treatment capability by providing patient health data to the physician

Fig. 2.10 Transportable Qardiocore intelligent transit healthcare EVS schema with IoMT technology

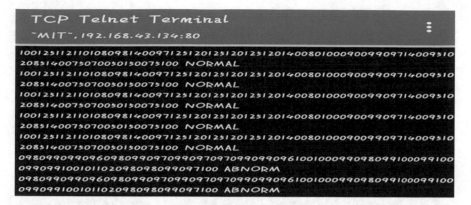

Fig. 2.11 ECG tracing and transmission to the IoMT database using a wi-fi module

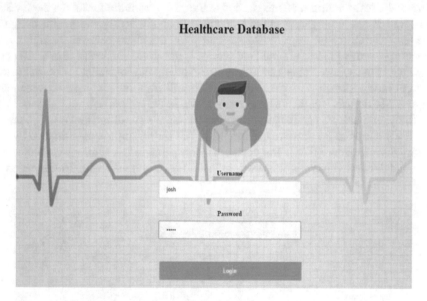

Fig. 2.12 Physician portal for remote patient monitoring using an ITH-IoMT system

long before the rescue vehicle reaches the medical clinic. The correspondence between the emergency clinic and the smart EVS is bidirectional. In this manner, tele-medical aid can be provided by appointing an attendant inside the emergency vehicle, thereby enabling immediate assistance to be given to the victim with the guidance of the physician through IoMT. ITH-IoMT allows the patient to be treated on-demand because the physician is aware of the patient's vital statistics before the rescue vehicle arrives at the emergency clinic for medical support. Consequently, time can be saved by completing important examinations before patients arrive at the medical clinic. The pre-treatment examination uses IoMT concepts by transmitting all important information to the cloud during the victim's transit to the emergency clinic in the EVS. Figure 2.12 shows the physician portal for these data and the password-protected secure database.

Predicated on the determination of the patient's need for urgent treatment by artificial intelligence (AI), the medicinal services framework pushes the health-monitored data to the appropriate medical specialist with the aim of rapid medical care. Figure 2.13 shows the specialist's administrator user interface, where the specialist can review the patient's health data and prepare to start medical treatment immediately upon the victim's arrival at the medical clinic.

2.7.4   Remotely Monitored Patient Medical Data for Treatment

The health information that was pushed to the cloud from the IoMT via the EVS is retrieved in the emergency clinic's database analysis. Real-time health data are made available to the appropriate specialist for treatment referral. A significant advantage of this ITH-IoMT medical service is that patient's health details are coupled with time stamps. This interlinked data can help the specialist to determine the current health situation of the patient, which will improve early-phase treatment. Figure 2.14 depicts a patient's heart rate by EVS in the specialist's administrator UI.

Body temperature can help to determine a patient's condition and level of severity. Body temperature decreases alarmingly after serious injury, as estimated by the Injury Severity Score (ISS). The decrease in temperature is a result of vasoconstriction and decreased oxygen transport to the tissues. Figure 2.15 shows the patient's

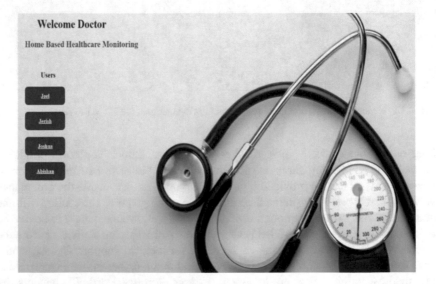

Fig. 2.13 Physician portal to view the patient's details

S.No	Heart Beat	Time
1	85	2017-04-17 18:05:09
2	95	2017-04-17 18:05:20
3	70	2017-04-17 18:05:25
4	70	2017-04-17 18:05:29
5	85	2017-04-17 18:05:42
6	90	2017-04-17 18:06:51
7	90	2017-04-17 18:07:06
8	87	2017-04-17 18:07:31
9	78	2017-04-17 18:07:49
10	82	2017-04-17 18:08:06

Fig. 2.14 A patient's continuously monitored heart rate

S.No	Body Temperature	Time
1	98.4	2017-04-17 18:59:55
2	98.9	2017-04-17 19:00:06

Fig. 2.15 Continuously monitored data for a patient's body temperature

variations in body temperature coupled with a time stamp, which is pushed to physician's portal to enable remote patient monitoring. By monitoring the patient's temperature, specialists can train the supervisor or janitor i.e. nurse with suitable counter activity for starting treatment and be prepared when the patient arrives at the emergency clinic. These preliminary screenings save valuable time and improve the patient's treatment process.

2.8 Conclusion

The ITH-IoMT schema described in this chapter is integrated with GPS-GSM to transmit automobile accident location details to the nearest emergency clinic for rapid treatment using a smart emergency vehicle system. The occurrence of an accident is discovered by a vehicle's vibration-detection sensor. A location alert is pushed to the cloud and the accident location is sent to a clinical database for rescue vehicle and medical services administration. The smart ambulance is equipped with a Qardiocore unit for ECG, respiratory rate, temperature, and pulse monitoring. Integrated medical devices can predict the patient's health condition when outfitted with IoMT innovations to enable remote patient monitoring. The remotely monitored health data is transmitted to the emergency clinic during the transport of the patient, which saves time by allowing the physician to review critical patient information long before the patient arrives at the clinic. This pre-treatment data can be retrieved with a specialist's referral to execute the essential course of action to treat the victim promptly after arriving at the emergency clinic. The transit time is used efficiently to improve patient outcomes with the support of ITH-IoMT ambulance services, which are equipped with ECG and body temperature monitors, Qardiocore, and other POC devices with IoMT integration for remote assistance between the victim and specialist, thus supporting the system of medical practitioners at the clinic long before the victim arrives.

References

1. Vishnu, S., Jino Ramson, S. R., & Jegan, R. (2020). *Internet of medical things (IoMT)-An overview*. In 2020 5th International Conference on Devices, Circuits and Systems (ICDCS), pp. 101–104. IEEE.
2. Catarinucci, L., De Donno, D., Mainetti, L., Palano, L., Patrono, L., Stefanizzi, M. L., & Tarricone, L. (2015). An IoT-aware architecture for smart healthcare systems. *IEEE Internet of Things Journal, 2*(6), 515–526.
3. Dumka, A., & Sah, A. (2019). Smart ambulance system using concept of big data and internet of things. In *Healthcare data analytics and management* (pp. 155–176). Academic Press.
4. Kumar, S. P., Akash, D., Murali, K., & Shriram, R.. (2016). *Call ambulance smart elderly monitoring system with nearest ambulance detection using android and Bluetooth*. In 2016 Second International Conference on Science Technology Engineering and Management (ICONSTEM), pp. 89–92. IEEE.
5. Sultana, T., & Patil, B. (2020). *Arduino based smart borewell ambulance rescue system*. In 2020 Second International Conference on Inventive Research in Computing Applications (ICIRCA), pp. 563–567. IEEE.
6. Loubet, G., Takacs, A., & Dragomirescu, D. (2019). Implementation of a battery-free wireless sensor for cyber-physical systems dedicated to structural health monitoring applications. *IEEE Access, 7*, 24679–24690.
7. Sanda, P. K., Barui, S., & Das, D. (2020). SMS Enabled smart vehicle tracking using GPS and GSM technologies: A cost-effective approach. In *Smart systems and IoT: Innovations in computing* (pp. 51–61). Singapore: Springer.
8. Fera, M. A., Aswini, R., Santhiya, M., Gayathiri Deepa, K. R., & Thangaprabha, M.. (2015). *HEAL-health monitoring in emergency vehicles with their authentication by RFID and location tracking by GPS*. In 2015 Seventh International Conference on Advanced Computing (ICoAC), pp. 1–6. IEEE.
9. Sudhindra, F., Annarao, S. J., Vani, R. M., Hunagund, P. V., & Enabled Real, A. G. S. M. (2014). Time simulated heart rate monitoring & control system. *International Journal of Research in Engineering and Technology, 3*, 6–10.
10. Mehta, M. (2015). ESP 8266: A breakthrough in wireless sensor networks and internet of things. *International Journal of Electronics and Communication Engineering & Technology, 6*(8), 7–11.
11. Bhate, S. V., Kulkarni, P. V., Lagad, S. D., Shinde, M. D., & Patil, S.. (2018). *IoT based intelligent traffic signal system for emergency vehicles*. In 2018 Second international conference on inventive communication and computational technologies (ICICCT), pp. 788–793. IEEE.
12. Thaker, T. (2016). *ESP8266 based implementation of wireless sensor network with Linux based web-server*. In 2016 Symposium on Colossal Data Analysis and Networking (CDAN), pp. 1–5. IEEE.

13. Sheth, A. (2016). Internet of things to smart IoT through semantic, cognitive, and perceptual computing. *IEEE Intelligent Systems, 31*(2), 108–112.
14. Al-Turjman, F., Nawaz, M. H., & Ulusar, U. D. (2020). Intelligence in the internet of medical things era: A systematic review of current and future trends. *Computer Communications, 150*, 644–660.
15. Wei, K., Zhang, L., Guo, Y., & Jiang, X. (2020). Health monitoring based on internet of medical things: Architecture, enabling technologies, and applications. *IEEE Access, 8*, 27468–27478.

Chapter 3
Protecting the Privacy of IoT-Based Health Records Using Blockchain Technology

Ahmet Ali Süzen and Burhan Duman

3.1 Introduction

The concept of the Internet of Things (IoT) has gained popularity and gained wide acceptance with the spread of the Internet and the fast improvement of information technology and smart devices. The IoT can be defined as having technological hardware capability and things having the Internet in communication with each other or with other systems. It also represents a network of interconnected objects or embedded devices with sensors over any kind network. [1]. In the IoT, objects must have a unique identity (unique addressability), be connected to the network, and generate data in order to be expressed as things. This makes things accessible and controllable from anywhere in the world [2].

IoT are very common areas of use such as health, energy, agriculture, IT, production, transportation, construction, and security. By utilizing IoT technologies, many applications are realized in these areas [3]. In applications, data generated by things is saved in local storage environments via the network or transferred to cloud storage media via connection to the Internet and covered in the cloud computing/big data context. Due to the fact that things have Internet connections, the data generated here also becomes available to the outside world, and data security issues/risks arise. In this case, the privacy of personal and commercial data may be lost.

The privacy and safety of patient data is of critical matter in the health sector, which is one of the most widely used areas. This personal data has patient privacy

A. Ali Süzen (✉)
Department of Computer Technologies, Isparta University of Applied Sciences,
Isparta, Turkey
e-mail: ahmetsuzen@isparta.edu.tr

B. Duman
Department of Computer Engineering, Isparta University of Applied Sciences, Isparta, Turkey
e-mail: burhanduman@isparta.edu.tr

© The Author(s), under exclusive license to Springer
Nature Switzerland AG 2021
D. J. Hemanth et al. (eds.), *Internet of Medical Things*, Internet of Things,
https://doi.org/10.1007/978-3-030-63937-2_3

and is an important issue within patient rights [4]. Medical data, also called electronic patient records, contain high levels of personal information such as patient identification information, medical history used treatment methods, dietary habits, and genetic information. Although the purpose of these records is to facilitate the management of diagnoses and treatments, they may also be used for different purposes by various institutions or individuals, such as hospitals and insurance companies. Privacy is not limited only to the storage and security of this data. In this context, issues such as preventing unauthorized individuals from accessing the system, keeping the source of health expenses a secret, obtaining the consent of the patient in advance if access to tertiary persons is to be granted, and ensuring a reasonable privacy environment are evaluated [5].

While there are multiple methods to ensure the confidentiality and safety of patient data produced with IoT, one of the methods that have been popular in recent years is the blockchain model. Blockchain is a technology that was announced by Satoshi Nakamoto on October 31, 2008, and forms the basic infrastructure of the cryptocurrency Bitcoin [6]. In very simple meaning, blockchain is the distribution of central trust on the Internet, which allows the removal of a central server or a trusted authority. Although blockchain technology is generally known as the technology beneath virtual currencies such as Bitcoin and Ethereum, it has a much wider range of possibilities and diversified applications. Blockchain is a decentralized encryption registry that allows the transfer of valuable assets in the current Internet environment that provides data transfer [7]. Blockchain technology is a significant factor in security since it is a structurally distributed system. Data is created, verified, and saved as encryption blocks with this technology [8]. Blockchain is the technology that has the ability to store encrypted data without the need for a reliable third party, so digital money transfers, contracts, banking, healthcare, voting, education. It is a technology suitable for use in different sectors such as intelligent systems, law, etc. [9].

Researchers have stated that while blockchain technology is a popular technology, there have been some problems and concerns because it is still a new technology [10, 11]. Data stored cryptographically encrypted on a blockchain network may have transaction histories reversed or access to data in the event that private keys in a permitted network are hacked by malicious persons [12, 13].

According to Deloitte (2016), hacking a patient's private key could limit possible negative damage, because the hacker will have to hack each user individually to acquire matchless private keys in accessing identifiable value information. The process of synchronized encryption will protect patient identities moving between or within organizations. Companies in the blockchain ecosystem can keep an updated ledger of their book of confidential data. If a block is to be regulated, security will be increased, and the risk of malicious activity will be reduced, as 51% of network users will need to confirm the change. Copies that are protected against malicious attacks will be provided at the time of the changes, thanks to network publication and distributed ledger [14].

Azaria et al., in their study on smart contracts, developed a method to access medical data using blockchain. In the model called MedRec, medical data is securely shared through blockchain. Instead of storing the record directly, MedRec encodes metadata that allows records to be securely accessed by patients and combines access to those given between different providers [15].

A medical data protection and sharing plan supported special blockchain were proposed by Liu et al. to improve the hospital digital health system. The program can meet numerous security features such as decentralization, clarity, and distraction resistance. A trustworthy operation has been established for doctors to hold medical data or reach patients' historical data also protecting privacy [16].

The model proposed by Attia et al. performs a remote healthcare monitoring of patients outside the hospital. For this purpose, the person's health status information (heart rate, blood pressure and oxygen saturation, body temperature, etc.) is obtained. In addition, different sensors can be installed to monitor the patient's immediate environment and to detect a number of events. The data broadcast by these sensors is permanently uploaded to a remote database system. It is clear that the system should have limited access and be secure, as these processes will be carried out over highly sensitive personal data. An architecture based on blockchain technology for remote monitoring of patient status has been developed to meet specified requirements. The architecture shown in Fig. 3.1 basically consists of two blockchains, a monitoring system and medical devices [17].

In a blockchain-based system proposed by Shen et al. for privacy-protected medical image retrieval, requirements were first stated in medical image retrieval and system design. Using blockchain techniques, the layered structure and threatening

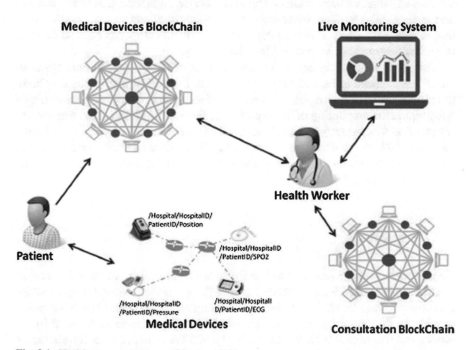

Fig. 3.1 Healthcare monitoring architecture [17]

model of the suggested system is represented. To place large medical images with storage restricted blocks, a customized processing structure is designed to protect the privacy of images and image features by removing a cautiously selected feature vector from the images [18].

Srivastava et al. have addressed the inclusion of blockchain technology in IoT based on health monitoring systems in his work. It has offered the benefits and hurdles of blockchain safety approaches in health monitoring using IoT devices. In addition, several potentially suitable cryptographic technologies have been evaluated for deployment in IoT [19].

Blockchain and IoT are a natural combination. Blockchain is a distributed technology, and IoT networks are often used to collect data from distributed sources [11]. The blockchain can ensure that the history and interactions of the processes performed by devices in IoT are kept in an unalterable manner. In future studies, devices will have the ability to process these records among themselves using them as a verification tool without the need for a central authority. Both devices can use mutual digital signatures and smart contracts to enable message exchange without adhering to a third authority. In addition to security features and cryptographic applications, blockchain technology has been providing an ideal platform for machine-to-machine operations [12].

In the studies reviewed so far, it is seen that the data set cannot be kept on IoT devices due to security problems, but this is now possible with the help of blockchain technology. Different IoT architectures are also proposed where data can be stored distributed without being collected in a center. Authentication in IoT devices and sending information to other devices in the ecosystem are seen as another problem in point of security. It is seen that blockchain architecture can be used for end-to-end messaging transactions and for data security.

This study aims at the secure communication of remote patient care systems, which are intended to be widely used in the near future. For the proposed system, the blockchain structure of pulse and galvanic sensor data connected to the patient's body allows the monitoring of the expert at the far point. Arduino Nano was used to retrieve data from sensors and send them to RAP. Raspberry PI was preferred for the data received to be included in the blockchain structure and transferred to Hyperledger in VPS. In the final step, the software has been developed for the expert to monitor the data.

3.2 Blockchain

Conceptually, a blockchain is a linked list in which each node in the list is named block. Records are added as transactions, and these transactions are brought to hash and clustered into blocks. Every block is cryptographically connected to the previous block, forming an ever-growing chain of blocks. Transactions stored in the blockchain (connected with a mathematical hash) are provided by a cryptographic method called the Merkle tree or hash tree. Figure 3.2 shows a general blockchain.

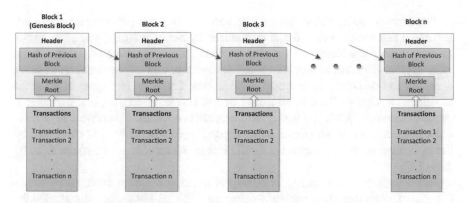

Fig. 3.2 Basic blockchain structure [22]

The first block is named the Genesis Block. This Block is known to blockchain participants and serves as a common reference. Each block after creation can contain several transactions and other metadata such as timestamp, block height (distance from Genesis Block). Once the data is written into a block, it cannot be easily changed, and the creation of a new block on the chain requires consensus (51%) among blockchain peers. Rollback of a transaction or deletion of data is logged as another transaction in a block, thus ensuring a strong data source guarantee (traceability). In a blockchain, all participants are spouses whose identities can be known or hidden [20, 21].

Blockchain networks can be categorized as permissive-unauthorized with their secret spouse identities and with their known spouse identities according to permit models that determine who can protect the network. If anyone can issue a new block, it is disallowed. If only certain users can publish blocks, it is called permission [23]. Unauthorized blockchains are more popular for cryptocurrencies, where any peer can join the blockchain network and participate in the consensus protocol. Permitted blockchains are more suitable for business applications and provide a different peer consensus path than unauthorized blockchains [24].

Blockchain business technology has also revealed new developments. One of these is smart contracts, first described by Szabo (1996) as "a set of promises stated in digital format, including the protocols that the parties will fulfill for promises" [25]. The use of smart contracts has become common with blockchain. Blockchain, a remarkable technique for the creation of smart contracts, is important as a record digital ledger system for any changes made to smart contracts or their contexts [26]. Smart contracts have some unique features when implemented on the blockchain. One of them is that the software code of a smart contract will be registered and confirmed, thus building the tamper-proof contract. Another is that the implementation of an intelligent contract is implemented between anonymous, insecure individual nodes without third party organizing and centriccontrol. Moreover, the smart contract can have its own cryptocurrencies or other digital entities, such as a smart broker, and transfer them after predefined terms are started [27].

Ethereum is the main blockchain platform that supports public smart contracts. Ethereum has been developed and customized by aid of a Turing-complete virtual machine named the Ethereum Virtual Machine (EVM) [28]. Ethereum differs from other cryptocurrencies in that it lets the improvement and implementation of smart contracts and delivered autonomous applications. There are two types of accounts in Ethereum. One of them is externally owned accounts (EOAs), and the other is contract accounts. While the first type of account is controlled by private keys without a code associated with them, in the second type of account, it is controlled by the associated code and contract codes. Transaction can only start through EOA by users [27].

In an analysis study on blockchain-specific smart contract applications established by Udokwu et al., it was shown that most (87.5%) of those who adopted the smart contract application were commercial. It is stated that 75% of the projects currently implemented are designed for special installations, plenty of the projects (62.5%) are prototyped or carried out, and just less than half are included by theoretical introductions and proposed circles/directions. Installments that adopt smart contract applications (71.87%) stated that it offers solutions in the areas of supply chain management (SCM), finance, healthcare, information security, smart city, and IoT (Fig. 3.3) [29].

The structures of distributed ledgers on which blockchain for IoT also has various features and applications. The distributed ledgers on the blockchain are Ethereum and Hyperledger Fabric. Another is IOTA (iota.org) based on the blockless distributed ledger architecture. Table 3.1 is given to show the usability of structures using distributed ledger architecture in IoT.

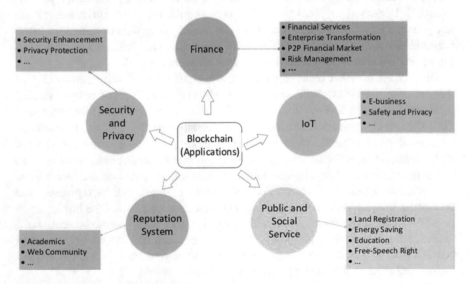

Fig. 3.3 Blockchain applications [29]

Table 3.1 Similarity of distributed technologies [30]

	Bitcoin	Ethereum	Hyperledger Fabric	IOTA
Native cryptocurrency	+	+	−	+
Decentralized app.	very limited	+ (Solidity)	+ (chaincode: Go, Java)	very limited
Transaction fee	v	+	−	−
Network type	Public	Public	Public (or private)	Public
Network access	+	+	−	Both
State channels	Lightning	μRaiden, Raiden	Not required	Not required
Suitable (IoT)	−	−	+	+
Suitable (Apss)	-	+	+	−

3.2.1 Blockchain Advantages and Challenges

Ethereum blockchain supports the intelligent contract and autonomous distributed applications, so it is widely used in other industrial applications such as management, IoT energy, and Health [27, 31–35].

Because of its many advantages, blockchain has found application space in sectors. The main advantages that make blockchain technology popular can be said to be security, decentralized structure, accuracy, cost, efficiency, privacy, transparency, traceability, and automation (smart contracts) [36–38].

Distribution: Blockchain does not store information centrally. The blockchain has been copied and distributed to a number of computational networks. When the block is created and attached to the blockchain, As soon as a new block is added to the blockchain, all nodes in the network update the blockchains to reflect the change. Thus by distributing knowledge to a network instead of depositing it in a single spot, it becomes more difficult to tamper with or lose the blockchain due to hardware failures. If a hacker gets a copy of the blockchain, only a single copy of the information will be compromised instead of the entire network [37].

Security: As soon as a transaction is recorded, a consensus must be reached that the authenticity of the recording must be confirmed by all nodes in the network. When the operation verifies, a block that contains all the relevant information is added to the blockchain. When knowledge in the chain is changed consciously or unconsciously, the hash code of that block changes, but the hash code on the blockchain does not change. This inconsistency makes it very hard for information about the blockchain to be altered or edited without prior attention [37, 39].

Accuracy: Blockchain operations are always approved by a great number of nodes on the network. This facilitates the verification process and provides a more accurate record of information by eliminating all human involvement, and human error is virtually nonexistent [36].

Cost: Actually, a blockchain is often more costly in dedicated facilities than a conventional centralized method. A blockchain, however, substitutes the need for reliable third-party mediator. Therefore, it lets saving on general expenses that are often expensive [37, 40].

Efficiency: Due to the great burden on the central authority, it can take several days to complete the operations performed through a central authority. In a non-working day check deposit transaction, the money may not actually appear in the account until the business hour of the business day. While financial institutions operate during normal business hours, the blockchain can run continuously 24 hours a day. Transactions are finished less time and can be thought safe [37].

Privacy: While blockchain networks are understood as anonymous, they are just confidential, not anonymous. So when a user does a transaction, their matchless code, named a public key, is saved on the blockchain instead of their individual data. Even if somebody's identity is till now connected with the blockchain address, the hacker cannot retrieve the user's personal information [37].

Transparency: Blockchain technology is almost always open source. Each user can see public blocks and processes stored within them. In the blockchain ecosystem, users can modify the code at will, as long as they have the majority of the network's processing power. Keeping the data on the blockchain makes it much more difficult to play with the data. Any user in the blockchain structure at a random time is unlikely to make an unnoticed change [37, 39].

Smart contracts: The main common blockchain feature is the ability to execute smart contracts. It allows for new capabilities that are not possible in a traditional system, with the ability to have a code that is predefined and works independently of participants' control [40]. Smart contracts are divided into two groups as public smart contracts and allowed smart contracts according to the blockchain platforms they operate on. Public smart contracts are used in healthcare and medical records, identity management, and scaling blockchains [41]. Permissioned smart contracts also are used in banking provenance and supply chain, IoT, insurance, and voting [42].

In addition to the many advantages that blockchain brings to traditional IoT applications, there are still many disadvantages and challenges in its application. The biggest problems arise from the limitations of IoT devices. Issues such as task distribution, power consumption, and computing capability need to be considered for blockchain to be effectively implemented in most IoT applications.

It appears that the energy costs associated with the required transaction capacity for public blockchain networks are high. Considering that as many devices will join the IoT ecosystem as time progresses, it seems that energy consumption will be a big problem.

When too many verification requests are made, the prolongation of the time required to process transactions also arises as a separate problem. Stronger hardware structures are needed in order to perform these processes in a shorter time.

It is known that data sets obtained from IoT devices are collected in a central device. The individuals or systems that produce the data may not want to share their data sets permanently. In other words, membership information is stored and shared by all members, including the center. But the lifetime of these data subjects is another problem that needs attention. Owners of data sets may not want to share them permanently. However, once any transaction is recorded by the blockchain, it cannot be changed or deleted. While this is a strong security property, if any record needs to be removed, it may not be suitable for sharing.

3.3   Proposed Model

The proposed safe patient monitoring model is shown in Fig. 3.4. Accordingly, the model comprises four parts. The general model consists of collecting patient data in the first part, transferring data to RAP in the second part, storing data in VPS in the third part, and monitoring software in the fourth part.

Fig. 3.4 Scheme of the proposed model

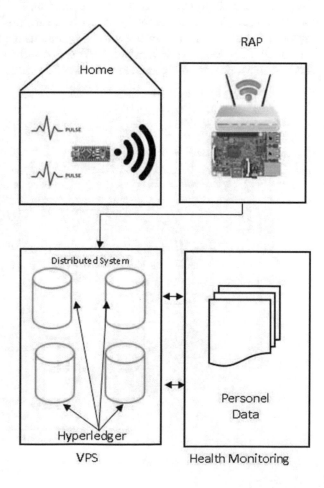

3.3.1 Personal Health Data Collection

Pulse and galvanic skin response sensors were used to obtain the data to be used in the proposed model. The control box with Arduino Nano is designed for instantaneous data retrieval from the patient. This control box allows data from the patient to be transferred to Raspberry Pi (RAP). The circuit diagram for the control box is shown in Fig. 3.5. The control box is transmitting the collected data to RAP via the ESP8266 Wi-Fi module.

3.3.2 Configuring Raspberry Pi Access Point (RAP)

A local wireless network is set up between RAP and health data from the patient for the patient monitoring system. In the proposed model, RAP is tasked with transferring data securely to the server. For this operation, Raspberry PI is used as four access points (AP). The Raspberry Pi hardware features the Raspbian operating system. The next section describes the configurations required to use the hardware as an AP. The "etc/network/interfaces" file was opened with the text editing tool, and the network was configured (Pseudocode 3.1) within the operating system. It

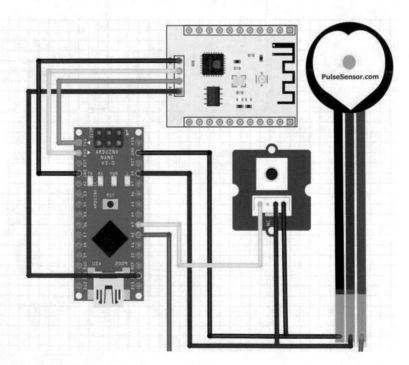

Fig. 3.5 Diagram of the control card

must also be changed to "net. ipv4.ip_forward=1" in order to use Raspberry Pi as an AP (/etc/sysctl.conf).

The "hostapd" service is used to broadcast a model-specific SSID. This allows them to connect of Arduino Nano to RAP with a password. To configure Hotspot settings, the "hostapd.conf" file is edited as in Pseudocode 3.2. In the last step, network address translation (NAT) must be set between Ethernet and Wi-Fi. This allows packets to be routed between Ethernet and Wi-Fi. The following commands are entered into the terminal to complete the process:

```
sudo iptables -t nat -A POSTROUTING -o eth0 -j MASQUERADE
sudo iptables -A FORWARD -i eth0 -o wlan0 -m state --state RELATED,
ESTABLISHED -j ACCEPT
sudo iptables -A FORWARD -i wlan0 -o eth0 -j ACCEPT
```

Pseudocode 3.1 etc/network/interface.d file configuration

```
auto lo
iface lo inet loopback
#eth0 settings option 1
#Uncomment next two lines to get available IP from local network
#auto eth0
#iface eth0 inet dhcp
#eth0 settings option 2
#Uncomment next five lines to get static IP from local network
auto eth0
iface eth0 inet static
    address 192.168.0.69
    netmask 255.255.255.0
    gateway 192.168.0.254
#Hotspot interface settings
allow-hotplug wlan0
iface wlan0 inet static
    address 192.168.X.1
    netmask 255.255.255.0
    network 192.168.X.0
    broadcast 192.168. X.255
```

Pseudocode 3.2 hostapd.conf file configuration

```
interface=wlan0
hw_mode=g
channel=11
wmm_enabled=1
macaddr_acl=0
auth_algs=1
ignore_broadcast_ssid=0
wpa=2
wpa_key_mgmt=WPA-PSK
wpa_pairwise=TKIP
ssid=Health_Test
wpa_passphrase=test123
```

Within the model, the data from the sensors are sent along with the device identification and time stamp. Pulse and galvanic measurement of the patient is performed in 5-minute periods as shown in Fig. 3.6, and 1-hour average info is forwarded to the server. The sample sent data format is

{c:1; Personal_Id:101 Pulse:91; Galvanic:448;17.07.2020 22.20}.

3.3.3 *Virtual Private Server (VPS)-Based Hyperledger Fabric Framework*

Hyperledger Fabric (HF) was used to store data sent from rap in a distributed blockchain structure. The HF system is preferred because of its data privacy and smart contact features [43]. The concept of smart contact is described as a programmable protocol [44]. Each action performed within the Hyperledger system is converted into smart contact. All smart contacts created within the model are called chain code. Ubuntu 18.04 operating system is installed to run HLF on VPS [42, 45].

Device_Id: 1	Personel_Id: 101	Pulse: 98	Galvanic: 420	Date: 17.07.2020 22:25

Add data

Device_Id: 1	Personel_Id: 101	Pulse: 96	Galvanic: 425	Date: 17.07.2020 22:10
Device_Id: 1	Personel_Id: 101	Pulse: 94	Galvanic: 435	Date: 17.07.2020 22:15
Device_Id: 1	Personel_Id: 101	Pulse: 91	Galvanic: 448	Date: 17.07.2020 22:20

...

Fig. 3.6 Collection of sensor data

Subnets called channels are created for the use of smart contracts within VPS. Each subnet that occurs has a separate ledger. Ledger is a registry that does not have features such as deleting and editing, which allows only the latest registry to be added. Only members of that group can see transactions performed on subnets. Each member has identifiers in the name identity that distinguish it from each other. Group privacy is ensured by channels. In this modeled HLF system, no one is allowed to access the platform, except for users who are authorized to access the network [46].

The platforms used on the HF system can be summarized as follows:

- API: Acts as a connector to the HF network by providing endpoints for chain functions
- Simulation: Simulates the functions of participant, manager, or controller
- Control panel: Displays different data according to user role in the system [47, 48]

It has been sending to HLF with JSON packages from RAP. The query.js file was developed with JavaScript language to display the saved data, and the process was carried out with "queryAllLogs" transaction. The transaction definition for sending data is as follows:

async queryAllLogs (device_id, personel_id, pulse, galvanic, DateTime) {...}

The editing method is automatically checked by *HealthDataContract* when the expert a request to show the data.

class HealthDataContract extends Contract {...}

So that, add new data to the chain, the invoke.js file encoded with Javascript has been developed and the process is completed with "createLog" transaction.

In order to keep the data hidden within the application, the client application keeps it as temporary data until the request is completed. The chain code generated during the process will be responsible for pulling temporary data and transferring it to the data store. The API used to perform this process is shown in Pseudocode 3.3. In addition, the structure of the Hyperledger fabric framework is shown in Fig. 3.7. Accordingly, 1c and 2c correspond to the process of adding and updating data in health1 and health2, respectively. If the State is destroyed for any reason, all operations in the blockchain can be recreated by repeating. According to the proposed structure, health1 cannot directly access health2. This is because our chain codes and smart contracts are designed to keep those chain codes and state separate from each other.

Devices from which data is obtained create an array (S1) containing the device header, server information, and data size to be sent to the AP. In the second step, the first array is combined with the array containing the health data (S2) to complete the task. In the Fabric section, a new transaction opens to include the incoming data in the chain (P1). Then the transport for the distributed architecture is created (P2), and the request is performed (P3).

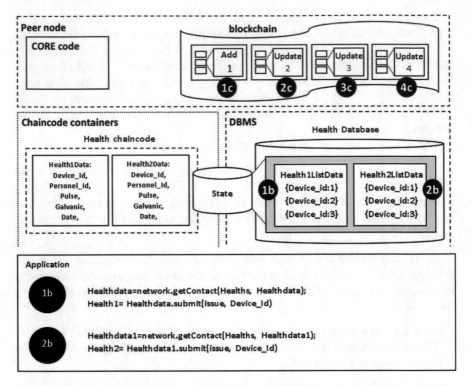

Fig. 3.7 Hyperledger fabric framework architecture

Pseudocode 3.3 Using fabric-network API and using client API

(S) // Private data sent as transient data:
(S1) const tempData = {
    sourcename: Buffer.from('Device1'),
    device: Buffer.from('Device1'),
    own: Buffer.from('Server'),
    value1: Buffer.from('800'),
    value2: Buffer.from('1200')};
(S2) const datalist =
await contract.createTransaction(
'initHealth)
    .setTransient(tempData)
    .submit();
	(P) // private data
	(P1) const txid
	    = client.newTransactionID();
	(P2) const sender = {
	    chain_code_Id : chain_code_Id,
	    tx-Id: tx_id,
	    fcn: 'initHealth',
	    args: [],
	    transientMap: temp_data };
	(P3)
	const endorsementlist= await channel.
	sendTransaction_Proposal(request);

The model is secured using TLS in all communication within the VPS and RAP. Certificate authority (CA) is required to connect to network components. For this, the CA certificates are stored in the directory containing the byfn.sh script.

CA Certificate for RAP1

cryptoconfig/peerOrganizations/rap1.example.com/peers/peer0.rap1.example.com/tls/ca.crt

3.3.4 Remote Monitoring Software Development

Interface software has been developed on the Visual Studio 2019 platform for remote monitoring of health data by experts with the C# programming language as seen in Fig. 3.8. This software securely demonstrates incoming data.

The recording of the patient to be monitored starts with finding the data holder from the VPS. The doctor makes a request from the software to view the data. If the patient who wants to see health data has authority, the private key is returned. Once the patient has access to their health records, they can also provide access to the query request, which they must store in their manager. In this way, he can access and decrypt the record in the database in the VPS. The process of monitoring and updating the data in the monitoring software is carried out in the following steps as visualized in Fig. 3.9.

- The expert sends a request to the administrator to view the data or add new data with the tracking software.
- The administrator sends a request to the blockchain to receive the patient's address and abstract key from the HealthDataContract.
- The application sends a request data to BC to create a new patient service and associate the person receiving the request with it.

Fig. 3.8 Health monitoring interface

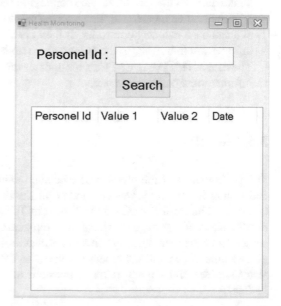

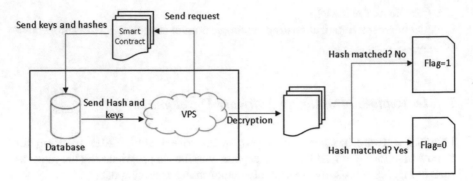

Fig. 3.9 Data receiving process

Table 3.2 Time for data saved and receive process from the model

Data type (block)	Avg. time (s)	File size (KB)
Creation	0.612	0.852
Updating	0.521	0.942
Sharing	0.520	0.852
Deletion	0.621	0.852

- Miners verify requests to BC and apply the task specified in the patient health data contract.
- The abstract of the contract is updated in the patient node and the patient is sent feedback on the arrangement through an authorized intermediary.
- Upon the conclusion of the response, a regulation in accordance with the patient health data contract in the service contract is sent.
- The chain node created for the patient sends a signed query request to the model's database gateway controller. It then checks permissions to determine what information to transfer to the patient chain node with the query.
- The data node's Database Protector updates the database of the patient node with information from patient node.

3.4   Results

The performance of the model was evaluated using JMeter. Blockchain modeling and testing in the study was conducted on the i9 CPU and Nvidia Inno3d 11GB hardware. The results obtained are shown in Table 3.2. According to performance results, block file storage, retrieval, and replacement in the cloud have been consumed less time than other systems in terms of cost, file size, and time. Figure 3.10 gives a time-dependent increment graph of the data coming to the model. File size also increases and latency remains constant at average values with the increase in data.

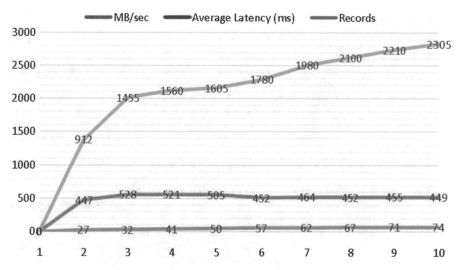

Fig. 3.10 performance evaluation of the model

3.5 Conclusions

In this paper, a blockchain-based approach has been recommended to securely transfer and monitor data in remote patient monitoring. For development of our model, pulse and galvanic electrical signal values, which are important data in determining the life risks of individuals, were used. The data is received from the sensors via the Arduino development card and sent to the RAP located on the local network. RAP sends the incoming data to the server by incorporating it into the given blockchain structure. All incoming records are distributed in the blockchain model thanks to software developed with the Hyperledger Fabric system running on VPS. Local interface software has been developed for experts to study patient data. This allows access to the data by requesting the requested data from the server.

The study was intended to be taken from given IoT devices and stored safely on remote health servers. In future studies, it considers evaluating the performance of the proposed system in its more scalable and realistic environments. Another planned study involves improving data security and privacy in the event of external access to private blockchain networks.

References

1. Khan, M. A., & Salah, K. (2018). IoT security: Review, blockchain solutions, and open challenges. *Future Generation Computer Systems, 82*, 395–411., ISSN 0167-739X. https://doi.org/10.1016/j.future.2017.11.022.
2. Wiki. (2020). https://tr.wikipedia.org/wiki/Nesnelerin_interneti
3. Gündüz, M. Z., & Daş, R. (2018). Nesnelerin interneti: Gelişimi, bileşenleri ve uygulama alanları. *Pamukkale University Journal of Engineering Sciences, 24*(2), 327–335.
4. İzgi, M. C. (2014). Mahremiyet kavramı bağlamında kişisel sağlık verileri. *Türkiye Biyoetik Dergisi, 1*(1), 25–37.
5. Çoban, Ç., & Tüysüz, M. F. (2019). E-health and privacy: Risks, opportunities and solutions. In *2019 4th International Conference on Computer Science and Engineering (UBMK), Samsun, Turkey* (pp. 554–559). https://doi.org/10.1109/UBMK.2019.8907079.
6. Nakamoto, S. (2008).Bitcoin: A peer-to-peer electronic cash system. Accessed 02 July 2010, https://git.dhimmel.com/bitcoin-whitepaper/
7. TUBİTAK, https://blokzincir.bilgem.tubitak.gov.tr/blok-zincir.html
8. Duman, B., & Şen, E. B. (2019). Blok zinciri Teknolojisi Bağlamında Nesnelerin İnterneti, Türkiye'de Mühendislik Alanında Yeni Yaklaşımlar, Editör: Dr. Serpil SAVCI, İKSAD, pp. 29–64, 138s, ISBN:978-605-7695-92-5.
9. Ghaffari, Z. (2016). *On the application areas of blockchain* (Master Thesis, Department of Computer Science, Malmö University), 33p.
10. Hileman, G., & Rauchs, M. (2017). *Global blockchain benchmarking study* (pp. 13–15). Cambridge: Cambridge Judge Business School.
11. Shafagh, H., Burkhalter, L., Hithnawi, A., & Duquennoy, S. (2017). Towards blockchain-based auditable storage and sharing of iot data. In *Proceedings of the 2017 on Cloud Computing Security Workshop* (Vol. 2017, pp. 45–50).
12. Hartman, W. T., Hansen, A., Vasquez, E., El-Tawab, S., & Altaii, K. (2018). Energy monitoring and control using Internet of Things (IoT) system. In *2018 Systems and Information Engineering Design Symposium (SIEDS)* (pp. 13–18). Charlottesville: IEEE.
13. Yavuz, M. S. (2019). Ekonomide Dijital Dönüşüm: Blockchain Teknolojisi ve Uygulama Alanları Üzerine Bir İnceleme. *Finans Ekonomi ve Sosyal Araştırmalar Dergisi, 4*(1), 15–29.
14. Deloitte, B. Opportunities for health care, https://www.healthit.gov/sites/default/files/4-37hhs_blockchain_challenge_deloitte_consulting_llp.pdf
15. Azaria, A., Ekblaw, A., Vieira, T., & Lippman, A. (2016). MedRec: Using blockchain for medical data access and permission management. In *IEEE 2nd International Conference on Open and Big Data* (pp. 25–30). https://doi.org/10.1109/OBD.2016.11.
16. Liu, X., Wang, Z., Jin, C., Li, F., & Li, G. (2019). A Blockchain-based medical data sharing and protection scheme. *IEEE Access, 7*, 118943–118953.
17. Attia, O., Khoufi, I., Laouiti, A., & Adjih, C. (2019). An IoT-Blockchain architecture based on hyperledger framework for healthcare monitoring application. *10th IFIP International Conference on New Technologies, Mobility and Security, NTMS 2019 – Proceedings and Workshop, 2019*, 1–5.
18. Shen, M., Deng, Y., Zhu, L., Du, X., & Guizani, N. (2019). Privacy-preserving image retrieval for medical IoT systems: A blockchain-based approach. *IEEE Network, 33*(5), 27–33.
19. Srivastava, S. S., Atre, M., Sharma, S., Gupta, R., & Shukla, S. K. (2019). Verity: Blockchains to detect insider attacks in DBMS. *arXiv preprint arXiv, 1901*, 00228.
20. Bonneau, J., Miller, A., Clark, J., Narayanan, A., Kroll, J. A., & Felten, E. W. (2015). SoK: Research perspectives and challenges for bitcoin and cryptocurrencies. *2015 IEEE Symposium on Security and Privacy, 2015*, 104–121.
21. Buterin, V. A next generation smart contract decentralized application platform. ethereum.org
22. Sultana, T., Almogren, A., Akbar, M., Zuair, M., Ullah, I., & Javaid, N. (2020). Data sharing system integrating access control mechanism using blockchain-based smart contracts for IoT devices. *Applied Sciences, 10*(2), 488.

23. Yaga, D., Peter, M., Nik, R., & Scarfone, K. (2018). *Blockchain technology overview* (National Institute of Standards and Technology, US Department of Commerce). Retrieved from https://arxiv.org/ftp/arxiv/papers/1906/1906.11078.pdf

24. Venkat, A. (2018). Blockchains and APIs, Accesed 25 July 2020 http://www.noahdatatech.com/blockchain-and-api/

25. Szabo, N. (1996). Smart contracts: Building blocks for digital markets. *EXTROPY: The Journal of Transhumanist Thought, 18*(2), 16. https://www.fon.hum.uva.nl/rob/Courses/InformationInSpeech/CDROM/Literature/LOTwinterschool2006/szabo.best.vwh.net/smart_contracts_2.html.

26. Zain, N. R. B. M., Ali, E. R. A. E., Abideen, A., & Rahman, H. A. (2019). Smart contract in blockchain: An exploration of legal framework in Malaysia. *Intellectual Discourse, 27*(2), 595–617.

27. Wang, S., Ouyang, L., Yuan, Y., Ni, X., Han, X., & Wang, F. (2019). Blockchain-enabled smart contracts: Architecture, applications, and future trends. *IEEE Transactions on Systems, Man, and Cybernetics: Systems, 49*(11), 2266–2277. https://doi.org/10.1109/TSMC.2019.2895123.

28. Rawat, R., Chougule, R., Singh, S., Dixit, S., & Kadam, G. B. P. A. (2019). Smart contracts using blockchain. *International Research Journal of Engineering and Technology (IRJET), 06*(06), 3880–3893.

29. Drosatos, G., & Kaldoudi, E. (2019). Blockchain applications in the biomedical domain: A scoping review. *Computational and Structural Biotechnology Journal, 17*, 229–240.

30. Pustišek, M., & Kos, A. (2018). Approaches to front-end IoT application development for the ethereum blockchain. *Procedia Computer Science, 129*, 410–419.

31. Liu, C. H., Lin, Q., & Wen, S. (2018). Blockchain-enabled data collection and sharing for industrial IoT with deep reinforcement learning. *IEEE Transactions on Industrial Informatics, 15*(6), 3516–3526.

32. Koirala, R. C., Dahal, K., & Matalonga, S. (2019). Supply chain using smart contract: A blockchain enabled model with traceability and ownership management. In *2019 9th International Conference on Cloud Computing, Data Science Engineering (Confluence)* (pp. 538–544). IEEE. https://doi.org/10.1109/CONFLUENCE.2019.8776900.

33. Longo, F., Nicoletti, L., Padovano, A., d'Atri, G., & Forte, M. (2019). Blockchain-enabled supply chain: An experimental study. *Computers Industrial Engineering, 136*, 57–69.

34. Khatoon, A. (2020). A blockchain-based smart contract system for healthcare management. *Electronics, 9*(1), 94.

35. Chang, S. E., Chen, Y., Lu, M., & Luo, H. L. (2020). Development and evaluation of a smart contract–Enabled blockchain system for home care service innovation: Mixed methods study. *JMIR Medical Informatics, 8*(7), e15472.

36. Gatteschi, V., Lamberti, F., & Demartini, C. (2019). An overview of blockchain-based applications for consumer electronics. In *2019 IEEE 23rd International Symposium on Consumer Technologies (ISCT)* (pp. 161–166). Wroclaw: IEEE.

37. Alam, M., Khan, I. R., & Tanweer, S. (2020). *Blockchain technology: A critical review and its proposed use in E-voting in India.* Available at SSRN 3570320.

38. Monrat, A. A., Schelén, O., & Andersson, K. (2019). A survey of blockchain from the perspectives of applications, challenges, and opportunities. *IEEE Access, 7*, 117134–117151.

39. Varghese, C. (2019). IoT device management using blockchain. *International Journal of Science, Engineering and Technology Research (IJSETR), 8*(3), 79–84.

40. Maesa, D. D. F., & Mori, P. (2020). Blockchain 3.0 applications survey. *Journal of Parallel and Distributed Computing, 138*, 99–114.

41. Shojaei, A., Flood, I., Moud, H. I., Hatami, M., & Zhang, X. (2019). An implementation of smart contracts by integrating BIM and blockchain. In *Proceedings of the Future Technologies Conference* (pp. 519–527). Cham: Springer.

42. Androulaki, E., Barger, A., Bortnikov, V., Cachin, C., Christidis, K., De Caro, A., & Muralidharan, S. (2018). Hyperledger fabric: A distributed operating system for permissioned blockchains. *Proceedings of the Thirteenth EuroSys Conference, 30*, 1–15.

43. Kumar, M. A., Radhesyam, V., & SrinivasaRao, B. (2019). Front-end IoT application for the bitcoin based on Proof of Elapsed Time (PoET). In *3rd International Conference on Inventive Systems and Control (ICISC)* (pp. 646–649). Coimbatore: IEEE.
44. Manevich, Y., Barger, A., & Tock, Y. (2019). Endorsement in Hyperledger Fabric via service discovery. *IBM Journal of Research and Development, 63*(2/3), 2–1.
45. Nawari, N. O. (2020). Blockchain technologies: Hyperledger fabric in BIM work processes. In *International Conference on Computing in Civil and Building Engineering* (pp. 813–823). Cham: Springer.
46. Benhamouda, F., Halevi, S., & Halevi, T. (2019). Supporting private data on hyperledger fabric with secure multiparty computation. *IBM Journal of Research and Development, 63*(2/3), 3–1.
47. Nasir, Q., Qasse, I. A., Abu Talib, M., & Nassif, A. B. (2018). Performance analysis of hyperledger fabric platforms. *Security and Communication Networks*. https://doi.org/10.1155/2018/3976093.
48. Goranović, A., Meisel, M., Wilker, S., & Sauter, T. (2019). Hyperledger fabric smart grid communication testbed on raspberry PI ARM architecture. In *International Workshop on Factory Communication Systems (WFCS)* (pp. 1–4). Sundsvall: IEEE.

Chapter 4
Medical Data Compression for Lossless Data Transmission and Archival

Ramesh Sekaran, Vimal Kumar Maaanuguru Nagaraju, Vijayalakshmi Jagadeesan, Manikandan Ramachandran, and Ambeshwar Kumar

4.1 Introduction to Medical Data Compression

In preceding years, medical imaging tests were performed over the radiological films. But, nowadays, medical imaging tests are performed digitally which improves the digital medical data quantitatively. Applications involving medical images generate a large and enormous amount of three-dimensional data. Moreover, in recent years, according to the electronic health record (EHR) systems, nearly 84% of hospitals are adopting digital medical data processing. This is because of the reason that HER stores' digital information like laboratory testing images, diagnosis results, clinical notes, and patients' prescriptions are linked with its treatment.

The propagation of the digital images, like computed tomography (CT), magnetic resonance imaging (MRI), electrocardiograph (ECG), ultrasound (US), and so on are broadening speedily. Storing such a vast amount of medical image data consumes a lot of overall disk space. Hence, there arises a requirement for efficient digital data storage and transmission.

Hence, compression is said to be the most prerequisite in medical applications. Moreover, such medical images have to be stockpiled without any information loss because of the reason that the image accuracy is sarcastic in diagnosis. This

R. Sekaran (✉)
Department of Information Technology, Velagapudi Ramakrishna Siddhartha Engineering College, Vijayawada, Andhra Pradesh, India

V. K. M. Nagaraju
Department of ECE, RMD Engineering College, Kavaraipettai, India
e-mail: mnv.ece@rmd.ac.in

V. Jagadeesan
Department of ECE, Kongu Engineering College, Perundurai, India

M. Ramachandran · A. Kumar
School of Computing, SASTRA Deemed University, Thanjavur, India

© The Author(s), under exclusive license to Springer
Nature Switzerland AG 2021
D. J. Hemanth et al. (eds.), *Internet of Medical Things*, Internet of Things,
https://doi.org/10.1007/978-3-030-63937-2_4

55

necessitates lossless compression techniques or also called as error-free compression with the decompressed image being similar to the actual image. Lossless image compression processes large digital image ensuring speedy interaction, ensuring quantitative data analysis. Also, the performance of the lossless compression is better than in a way they handle bandwidth management constraints and stockpiles.

4.1.1 Preliminaries

Contemporary developments in the area of information technology emerged to the origination of a vast amount of data at every fraction of seconds. Due to this, the data storage and transmission of data are probably to expand exponentially to a vast amount of data. Based on Parkinson's first law, the requirement of data storage and transmission expands at least two times its data storage and transmission potentialities.

With the availability of optical fibers, asymmetric digital subscriber line (ADSL), and an Ethernet cable, in recent years, a surge has occurred in the data growth rate when compared to the technology growth rate. Hence, it is found to compromise the abovesaid large data storage and transmission. To overcome this challenging issue, a substitute conception called DC has found its place in recent years to compress the data size storage and transmission. It reshapes the actual data to its compressed or condensed form by the identification and application of patterns that exist in the digital data.

The advancement of DC initiated with the Morse code, initiated by Samuel Morse in the year 1838 with the compression of letters in telegraphs. Samuel Morse's Morse code utilized compact sequences to denote usually materializing letters and therefore reducing the size of the message and time involved in transmission.

In recent years, DC techniques are crucial in most of the applications involving real-time system; to name a few are:

- Geographical information systems (GIS)
- Graphics
- Wireless sensor networks (WSN)
- Cloud computing (CC) and so on

Even though the image data quality is extremely increasing with the recent advancement in technology, data size is found to be increased. Therefore the limiting of image file size empowers in storing a higher amount of information in the indistinguishable storage area with a minimum amount of transmission time. Hence, without the utilization of DC, it is said to be very time-consuming and cumbersome to store or transfer enormous data files.

4.1.2   Definition

As innumerable DC methods have been designed, a requirement comes to light to assess the methods and roughly choose the algorithm to utilize in a specific circumstance. Simultaneously, it is very tremendous to categorize these methods under certain categorizations. In this section, the definition of DC techniques according to three different classifications, namely, standard [1], data classes, and real-time applications, is presented. The classification ranking is illustrated in the figure (Fig. 4.1).

4.1.2.1   According to Standards

Normally, a DC technique has control over the data standard to a definite magnitude based on the application in concern. Whenever a DC technique is utilized for prevalent grounds, such as sending and receiving messages, to name a few, restored data quality is not tremendously in consideration. However, in the case of compressing the text involved, the changes made in a single character concerning text document have a profound impact on the complete interpretation.

Similarly, when medical imaging is concerned, a bit of change made in the pixels makes a magnificent change and is hence not beneficial. Therefore the importance of the quality of data of a DC heavily is influenced by the application taken for consideration. DC is used to minimize storage deficiency and energy consumption. An effective finite impulse response filter was designed in [2] by the meta-heuristic techniques – a game theory-based approach. This approach is minimizing the noise

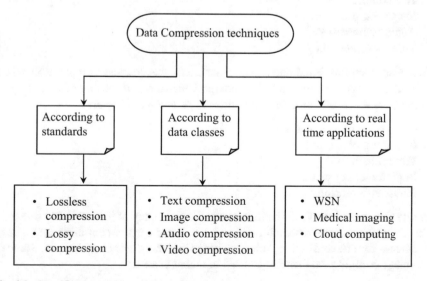

Fig. 4.1 Classification ranking of DC techniques

for enhancing the quality of the images. According to the necessity of reconstructed data standard, DC techniques are said to be split into two types. They are:

- Lossless compression
- Lossy compression

Lossless Compression Constitutes no information loss according to the name. In other words, the recreated image data is said to be similar to the actual image data. It is utilized in certain applications where the information loss is not acceptable. To name a few examples are:

- Text
- Medical image representation
- Satellite image surveillance

Lossy Compression From the name itself, a certain amount of information is said to be lost in case of lossy compression. In other words, in certain applications, the restructured data need not be the same as the actual data, and estimation of the actual image data is said to be acceptable. To name a few examples are:

- Joint Photographic Experts Group (JPEG) image
- Moving Picture Experts Group (MPEG) video

4.1.2.2 According to Data Classes

Normally, DC has been utilized for compressing, text, image, audio, and video. An overall review is described below.

- Text compression
- Image compression
- Audio compression
- Video compression

Text Compression Fundamentally, experiences that lossless compression where information loss is not permitted. A comprehensive preprocessing of text DC performed. Here consolidation of five entities is performed. They are:

- Conversion of a capital letter
- Inspecting of end of line
- Word substitution
- Replacement of words
- Phrase substitution

Image Compression With the increase in the frequency of images being transmitted day by day, several research persons have started carrying out different types of researches for efficient and significant storage and transmission of images. Some of the image compression techniques used in recent years are:

- Discrete cosine transform
- Huffman chain code
- Evolutionary programming format
- Chinese remainder theorem
- Prediction algorithm
- Singular value decomposition

Audio Compression Audio compression involves the procedure of minimizing a signal's range in a dynamic format. Dynamic format ranging refers to the differentiation between the heavy and noiseless portions of an audio signal taken for consideration. Different types of compressors are utilized by weakening the heaviest parts of the signal and increasing the result, so the noiseless portions are more obvious. A few types of audio compression techniques are:

- Audio attention
- Virtual reality content
- Huffman shift coding
- Intrinsic functions

Video Compression A video [3] is also an indispensable portion of multimedia applications. More often than not, video files monopolize a higher amount of resources while communicating, processing, and storage. Hence, compression is highly required and found to be essential for video file storage, video file processing, or video transmission. Various strategies were designed to significantly compress the video files to circumvent enormous data being transmitted or stored. A few examples are:

- MPEG – 1
- MPEG – 2
- MPEG – 3
- MPEG – 4

4.1.2.3   According to Real-Time Applications

Even though certain frequent compression methods are said to be appropriate to selected applications, the focus has been on method alternative to the application involved. Nevertheless, there are some methods where it is impractical to disconnect the method from the application. This is due to the reason that different methods are said to be dependent on the classification of data entangled in the application. A few examples are:

- Data compression in MANET/WANET
- Data prediction
- Wavelet transform
- Medical data compression in MANET/WANET
- Digital watermarking

• Satellite/aerial image

4.1.3 Significance of Medical Data Compression

DC has a predominant character in different types of data management. Minimizing the storage space indispensable without overlooking any information reduces the cost involved in the storage of specific data without compromising any empirical aspect. Small-scale files minimize the bandwidth demanded to send and receive data while operating more efficiently via I/O congestions to accelerate the data processing.

Certain data files hitherto utilize GZIP compression to minimize the proportions of these data files. Nevertheless, universal-purpose compression-like GZIP does not bestow optimal compression of specific data, and there is a substantial proportion of storage time and capacity being dissipated. The application of state-of-the-art revamped compression methods will minimize this dissipation.

4.1.4 Benefits of Medical Data Compression

Digital images are immensely data accelerated and therefore necessitate huge memory consumption and are also found to be time-consuming processes during transmission. With the aid of lossless image compression models, it is probable to eliminate certain unnecessary information present in images and therefore necessitate minimum storage and time consumed for transmission. Figure 4.2 shows a block diagram of the lossless image data compression process.

As shown in the above figure, the image compression comprises a lossless compression model and a lossless decompression model. In the figure, ' ' represents the actual image, and ' I ' represents the decompressed image. The size of the compressed image ' $n * n$ ' is lesser than the actual image size ' $m * m$ '. The main advantages of compression are as follows:

• Curtailment in data storage space, time, and bandwidth
• Cost reductions
• Reduces storage expenses
• Requires minimum amount of transfer time and lesser bandwidth
• Increases productivity

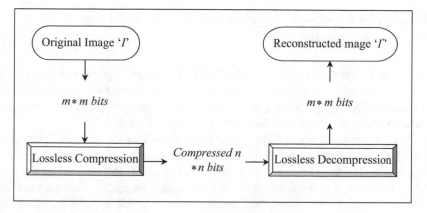

Fig. 4.2 Block diagram of lossless image compression/decompression

4.2   Designing and Coding for Lossless Data Transmission

Medical imaging has become an essential instrument in clinical application. Examinations have manifested correlations between the utilization of medical imaging testing and reduction of the rate of mortality to a greater extent and minimized the requirement for analytic surgery, lesser hospital stay, and so on. Due to this, the use of medical imaging has increased in a swift manner in the course of the preliminary segment of the last few years.

Digital image processing validates effortless retrieval of images, effective storage of image data, fast transmission of images for remote diagnoses, and the preservation of large image files for research purposes. To utilize techniques involving digital signaling, analog signals like X-rays have to be transformed into a digital type of format. During the process involving sampling, information loss is said to take place, therefore resulting in signal deterioration.

Large-scale medical analysis utilizing digitized data format necessitates a higher amount of transmission time, therefore resulting in issues related to data management, leading to confined disk storage issues. Therefore, only advanced technologies are not sufficient for image data transmission and storage. Designing and coding for lossless data transmission are advantageous and frequently crucial for cost and time saving involving data storage and transmission. On the other hand, decompression involves the inverse operation. OppNets was introduced in [4] to broadcast medical data in mHealth applications for enhancing the transmission performance. A new lossless data compression algorithm named adaptive lossless data compression (ALDC) algorithm was presented in [5] for wireless sensor networks to minimize the data transmission.

4.2.1 The Contemporary Prototype of Medical Data Compression

The contemporary prototype of the medical DC system comprises the compression processor, a host computer, and the reconstruction displays. First, the input images are acquired from any camera, or scanning devices are processed using the compression processor. Figure 4.3 shows the prototype of medical DC.

As depicted in the above figure, the output is the compressed data which is transmitted via a local area network (LAN) to the host computer, and the compressed data is stored in the database. Upon the request made by the user, the compressed data is transmitted to the display unit via a high-speed channel by the host computer. Finally, the reconstructed and displayed data is sent to the intended recipient with the aid of the built-in decoder.

4.2.2 Characteristics of Data Acquisition and Storage

The digital image in lossless data transmission comprises picture elements, also called pixels. Each pixel is connected to a number also referred to as the digital number. This digital number refers to the average luminosity of comparatively small regions within a frame. Based on the characteristics of high-speed data acquisition and storage [6], it becomes more effective and significant for the implementation of hardware in an easier manner that in turn also minimizes the frequency of data being collected. In this way, it minimizes the coercion on medical image data storage and transmission.

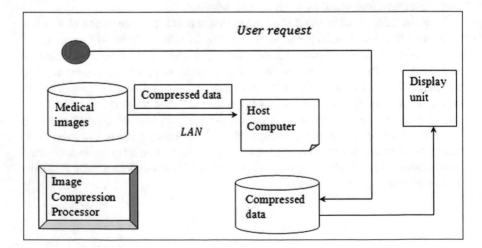

Fig. 4.3 Prototype of medical data compression

4.2.2.1 Data Acquisition and Characterization

To acquire data, conventional data transmission employs a simple and systematic sampling model using a matrix. This type of representation is found to be principally effective. The conventional sampling model for data acquisition however does not certainly counterpart the image information contents. Some of the data acquisition models are:

- Downsampling and decimation
- Upsampling and interpolation

Downsampling and Decimation

Decimation involves the procedure of filtering and downsampling a signal to minimize its efficient and significant rate involved in sampling. In other words, to downsample a signal refers to the process of obtaining a new signal of all sample of the actual signal. Here, filtering is used to put a stop to epithets that vigorously result from downsampling.

Upsampling and Interpolation

The procedure of shooting up the rate of the sample is referred to as interpolation. In the process of interpolation first, an upsample is applied, and then suitable filtering is used forming finally interpolation.

4.2.2.2 Data Storage and Characterization

Data storage refers to the collaborative models and techniques that extract and maintain digital information on different types of storage media. As far as digital devices are concerned, storage is said to be an essential element. This is because of the reason that both the consumers and producers have started depending on it to conserve the information stretching conventional photos to critical management information system structures.

Organizations now have started utilizing tiered storage to computerize data positioning on distinct storage media based on an application's dimension, consent, and production specification. Data storage is characterized into two types. They are:

- Primary storage
- Secondary storage

Primary Storage

Primary storage controls application tasks interior to an establishment's production concerning foremost business lines. In other words, primary storage is also referred to as the main storage or primary memory. In the case of primary storage, data is gripped in random access memory (RAM). Also, primary storage broadly bestows fastened access than secondary storage. This is due to the closeness of storage to the computer processor.

Secondary Storage

Secondary storage fringes data on different types of devices like a hard disk drive, magnetic tape/disk, and certain other devices necessitating I/O operations. In other words, secondary storage media is frequently utilized for cloud storage and holding backups. Also, secondary storage can bear much more data when compared to primary storage.

4.2.3   Designing

A thorough understanding of medical DC is required and is said to be acquired via design. The three most prerequisites for designing [7] of medical DC are:

- Data minimum in size consumes lesser transmission time, and hence the entire medical image data is said to arrive at the receiving end in a swift manner. This is said to be advantageous for both the transmitter and receiver due to the reason that it discharges the network dimensions for certain other reasons and minimize the issues related to the network. As far as high-volume transmitters are concerned, it is economically significant by only sending half the bytes of information.
- Utilization of network resources economically is better for all the users using an image as data those who share the network internal resources. Liter resources refer to more messages which can be fit in within the network's resource limitations.
- It is better to arrange in repletion with the precise objective of rectifying bit errors, in place of whatever suboptimal repetitions occur to existence in the actual message.

4.2.4   Entropy Coding

One of the types of lossless coding is entropy coding. This entropy coding compresses the given input digital data with the representation of commonly materializing patterns with minimum bits and seldom transpiring patterns with several bits. The entropy coding process is divided into two modules. They are modeling and coding. The process of modeling involves the efficient assignment of potentialities to the corresponding symbols. On the other hand, the process of coding generates sequences of bits from these potentialities. A most preferable entropy coding is Huffman coding.

The most frequently used lossless DC with an optimal prefix code is Huffman code. The procedure of identifying or processing such a code yields with the aid of Huffman coding. The output arrived from Huffman coding is said to be exhibited as a floating distance code table for encoding a character in a file. The algorithm generates this table from the approximated potentiality or density of weight for each probable value of the source data.

4.3   Data Compression Techniques for Lossless Data Transmission and Archival

The significance of data/image compression escalates with improving communication technology. The amount of data involved in transmission and archival linked to the visual type of information is found to be so large that its storage necessitates immeasurable storage potentiality. Data storage and transmission of such data necessitate comprehensive dimensions and bandwidth that could be very extravagant. Hence, data/image compression techniques are apprehensive with minimization of the number of bits essential for data storage or data transmission without any considerable information loss.

4.3.1   Coding Scheme

High compression rate and security aspects are said to be attained using coding. Some of the frequent techniques found under lossless classification [8] are given below.

- Run-length encoding (RLE)
- Huffman coding
- Lempel-Ziv-Welch (LZW) coding
- Area coding

4.3.1.1 Run-Length Encoding

One of the simplest forms of image compression technique is RLE that functions by computing the number of neighboring pixels with similar gray-level values. This count also referred to as the run length is coded and stored in the storage. The total number of bits utilized for coding in RLD depends on the pixels present in the row.

4.3.1.2 Huffman Coding

Huffman coding is a statistical lossless coding technique in [9]. As far as Huffman coding is concerned, it is said to produce a code that is as near as possible to the entropy, therefore emerging into variable-length coding. It is a lossless DC designed based on the ASCII characters that emerged by David Huffman in 1951. It involves a type of entropy coding mechanism where it transforms the finite length codes of characters to dynamic length codes of characters via nonrepeating prefix codes designed in such a manner that no two characters are allocated with similar codes.

4.3.1.3 Lempel-Ziv-Welch (LZW) Coding

LZW coding is said to be a comprehensive lossless DC algorithm developed by Abraham Lempel, Jacob Ziv, and Terry Welch in 1984. LZW coding is said to be a dictionary-based coding in the form of either a static or dynamic structure. In the case of static dictionary-based coding, during the process of encoding and decoding, the dictionary is said to be fixed, and on the other hand, in the case of dynamic dictionary-based coding, the dictionary is rationalized during processing. With this, the redundancy is said to be removed by ignoring tedious dictionary recommendations and hence is utilized in compression image types like GIF and BMPs.

4.3.1.4 Area Coding

The upgraded form of run-length coding is area coding that contemplates the two-dimensional image characteristics. The concept beyond this is to identify the rectangular portions with similar features. These rectangular portions are coded in an illustrative form as an ingredient in a pictorial form as an ingredient with dual points and a definite composition.

4.3.2   Data Quality Specification

Broadly, a lossless image compression technique impacts the data quality specification to a certain scope based on the application benchmark. As far as compression techniques are used for messaging, the data quality of regenerated data is not, particularly, in concern. However, in case of compression involving text, moderation even in the case of a single character is not admissible as it coppers the complete definition. Similarly, with medical images in concern, a small alteration make many changes. The importance of the data quality of the DC technique is tremendously dependent on the data type.

4.3.3   Application-Oriented Specification

Even though certain compression models are said to be applied to only certain applications, the focus has been on the technique utilized when compared to the application in vogue. In the case of wireless sensor networks, due to the energy-constrained or energy-limited nodes, the type of compression method utilized plays a pivotal role in reducing the energy being consumed. Here, the DC methods are utilized in reducing the data frequency and hence minimizing the frequency of data being transmitted to conserve energy or improve energy efficiency. A large amount of medical images and patient data are controlled and supervised in several hospitals and medical research institutes. More of the storage space and time for transmission is said to be consumed. Here, compression techniques are utilized for compressing the medical data and specifically found its application in telemedicine.

4.3.4   Bandwidth, Storage, and Data Compression Techniques

A straightforward classification of DC is that it necessitates reconstructing a string of characters into a new string which accommodates similar information but whose extent is as inconsiderable as possible. This is because of the reason that DC has the foremost application in the domain of data transmission and storage. Several data processing utilizations necessitate enormous data volume storage, and the frequency of those applications is continuously enlarging as the utilization of computers widens to state-of-the-art directions.

Similarly, the propagation of computer communication is emanating in tremendous data transfer over communication links. By compressing the data to be either stored or to be transmitted to the other end minimizes the storage being consumed and the communication cost involved in transmission also. When the frequency of data to be sent to the other end is minimized, the consequence is that of escalating the magnitude of the communication medium. Similarly, file compression to half of its actual size is analogous to magnifying the extent of the storage means.

4.4   Comprehensive Algorithms for Medical Data Compression

The purpose of medical DC is to minimize the size of medical data to store or transmit the data to the other end in an effective manner. Digital imaging and communications in medicine refer to one of the medical calibers to construct and perpetuate international qualities for transmission of information about biomedical diagnostics and therapeutics. It utilizes images corresponding to digital format and its corresponding data. Three principal concerns are to be taken into consideration with the DICOM communications. They are:

- Memory prerequisite
- Bandwidth limitation
- Batter resource curtailment

Some of the areas utilizing DICOM standards are:

- Endoscopy
- Mammography
- Pathology
- Surgery
- Dentistry
- Cardiology

The main objective of DICOM standard remains in enhancing productivity, effectiveness between imaging, and other information systems about healthcare environments globally.

4.4.1   Dictionary Past and Future Component Analytics

Conventional medical imaging initially requires to scan the portions of definite human regions and then acquire the respective image data and finally return the portion segment that in turn are sent for screening to the health professionals for further observation. In day-to-day life, human beings are amicable with the subsequent medical data images, like:

- Computed tomography
- Positron emission tomography
- Magnetic resonance imaging
- Ultrasonic imaging

Image data accommodate much dispensable information, and with the performance of medical image compression, irrelevant information is said to be removed. However, the redundancy of medical data includes:

- Spatial redundancy
- Temporal redundancy
- Time redundancy
- Frequency redundancy

Principal component analysis (PCA) [10] is here utilized in medical DC to predict high-dimensional data to low-dimensional space. Here, PCA compression is said to be performed on the respective medical image data, performs PCA processing via image matrix and organization of principal components, and finally generates the medical data image matrix compression ratio and contribution rate.

In the abstraction procedure of practical concerns, there customarily prevails determined dependency between variables. Hence, to utilize this feature to perform PCA not only minimizes the frequency of variables and also enhances the working effectiveness of the method but also reduces the prerequisites on certain items like hardware organization and improves the practicability of abstraction.

The necessary of PCA is to attempt to re-alter the actual variables into a classification of state-of-the-art, collaborative unrelated variables and, in consonance with the original requirements, obtain a considerable integrated variable to contemplate as much information of the actual variables as possible. Therefore, it is said to be an arithmetic model, also a statistical model for reducing the dimension. The fundamental conception of PCA is therefore to reconstruct several indices that possess definite association into a kind of new but communally unrelated consolidated indications and restore the actual ones.

PCA is an analytical model to identify the principal features of a sparse dataset based on the total deviation. Given a set of invariable sparse data in ' $-B$ ' coordinate system as illustrated in Fig. 4.4, PCA initially identifies the greatest discrepancies of the actual datasets.

As depicted in the above figure, the data points are extrapolated onto a new axis called ' $-Q$ ' coordinate model. The order of ' ' and ' Q ' axis is referred to as the principal components. The principal order in which the image data differs is reflected by ' ' axis succeeding by its extraneous order, ' Q ' axis. However, in the case where all the medical image data points on ' ' axis are very nearer to zero as illustrated in Fig. 4.4(b), the medical image data is said to be represented by only a single variable ' P ', and the variable ' Q ' is said to be eliminated.

4.4.2   Unbounded Context Partial Matching

In medical image lossless compression, the medical image pixels are concisely denoted in such a manner that it can be reformulated without any error rate. This is significant in certain applications for medical reasons where the images have to be

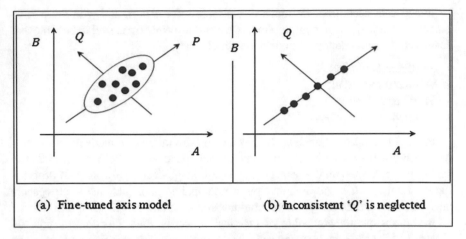

(a) Fine-tuned axis model   (b) Inconsistent 'Q' is neglected

Fig. 4.4 Principal component analysis for dimension reduction in medical images

stockpiled or carried out in different locations in such a manner that they can be reconstructed without any alterations in the medical image data. Lossless medical image compression models either consider the image for analysis in the form of a single dimension or make utilization of the two-dimensional contexts to enhance the performance of coding in a significant manner. The preceding model necessitates the earliest stage of two-dimension medical images to single-dimension medical images and a text compression technique to compress the single-dimension sequence.

Several of the fortunate lossless image compression techniques are based on the contextual pattern, and they utilize the two-dimension spatial redundancy in medical image data. Some of the examples are:

- Lossless Joint Photographic Experts Group (LJPEG)
- Fast, Efficient, and Lossless Image Compression System (FELICS)
- Context-based, adaptive, lossless image codec (CALIC)
- Set partitioning in hierarchical trees (SPIHT)

The abovesaid examples comprise four elements. They are:

- A primary prediction model to eliminate spatial redundancy between adjacent pixels
- A background alternative model for an image location
- Evaluation of the conditional possibility of occurrence of prediction error
- Entropy designed according to the evaluated conditional possibility of occurrence

Unbounded context partial matching for two-dimension spatial redundancy forms a predominant step in high-performance lossless DC. To efficiently determine and make use of contexts for medical images is however a grueling issue. This is specifically due to a large number of contexts in prevalence in medical data images, which generally results in the elevated cost involved in modeling resulting in mini-mized compression effectiveness. Here, partial matching, a method for context modeling of medical data images and compression, is utilized.

4.4.3   Positive Markov State-Space Transitions

The positive Markov state-space transitions have three compression states as given below. They are:

- Initialization and refresh state
- First-order state
- Second-order state

4.4.3.1   Initialization and Refresh State

In initialization and refresh state, the medical image data has been compressed using any compression algorithms, and information about the actual data is sent.

4.4.3.2   First-Order State

In the first-order state, the compression mechanism has been detected with a medi-cal data image in the receiving end and stores the fields like the IP address of the sender and port numbers of both the sender and receiver for establishing a connec-tion between two sides.

4.4.3.3   Second-Order State

In the case of the second-order state, certain dynamic variables like sequence num-bers are said to be subdued by sending only a sequence number of logical order and also a dynamic checksum to be verified on the other side.

As given above, each compression state utilizes a different header to send the information about the header to the corresponding receiver. The context of the med-ical data is said to be established in the IR state that includes both the static and dynamic information about the header. On the other hand, the FO state bestows the difference pattern involving dynamic fields. Finally, the SO state sends the com-pressed values like sequence number and timestamp being sent. Figure 4.5 shows the state-space transitions for lossless medical image compression.

To enhance the compression level, a positive model is utilized during lossless image compression to enable that the context has been precisely settled at the decompression end. This refers to the lossless image compression which utilizes the

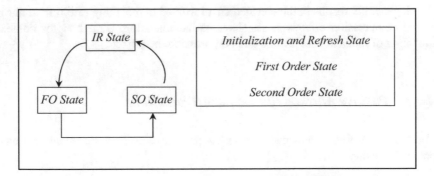

Fig. 4.5 State-space representations

same header format for different numbers of medical images. On the other hand, as the compression mechanism does not know the concept of context loss, two timers are utilized to further get back to the FO and IR states. Finally, the decompression mechanism works at the other end and further decompresses based on the information present in the header field of the context. Similar contexts are said to be utilized both in the compression side and decompression side to store information to make certain precise synchronization.

4.4.4 Binary Decision Tree Sequence

To reduce the lossless image compression storage and transmission, the requirement for image compression has significantly risen extremely. The binary decision tree sequence is aimed at sequentially improving the compression rate. Here, the binary decision tree is represented in the form of two probable values for each pixel. They are:

- Possessing 0 value
- Possessing 1 value

The binary decision tree by the name itself directs in two forms on the right side and the left side. In other words, the root tree or the actual image is split into two portions, either 0 value in the left/right side or the 1 value in the left/right side of the tree. Therefore it is called a binary decision tree. Here, the decision regarding the left or the right side representation is done according to the presence of pixel values in the image for consideration.

The binary decision tree sequence refers to a structure designed based on a sequential decision-making process. The procedure is said to be initiated from the root, followed by which a feature in the lossless image is evaluated and one of the two branches is handpicked. This process is said to be iterated until a final leaf is arrived at that usually denotes the compressed/decompressed image for analysis. The binary decision tree visualization should spotlight the following constituents:

- Feature target space
- Feature and feature split value
- Leaf node purity
- Leaf node prediction
- Samples involved
- Sample leaf node

Binary decision tree sequence or tree-based sequence searching is a very significant and computationally efficient method for lossless image compression [11]. The workflow of the binary decision tree sequence is summarized as shown in Figure 4.6.

As shown in the above figure, the process is said to be initiated with the splitting of the actual medical image also referred to as the original tree into two levels by splitting the medical image iteratively into four sub-images. At the first level, the actual medical image is split into four pieces, and at the second level, the image pieces are additionally split into four pieces. In this manner, an entire medical image is said to be denoted in the form of a tree. Next, the process of segmentation is conducted iteratively and examined at each iteration. In this way, the binary tree structures are fabricated by recurrent splits of the medical image.

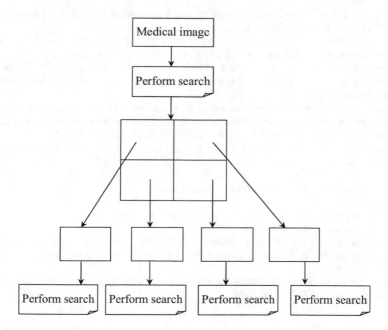

Fig. 4.6 Binary decision tree

4.5 Conclusion

In this chapter, an overview of data compression techniques carried out by previous researches is presented. DC techniques are used for three different classifications, namely, standard, data classes, and real-time applications, which are explained. The coding and designing are used for cost and time saving involving data storage and transmission. Many different DC techniques on three classifications, namely, coding schemes, data quality specifications, and application appropriateness, were analyzed. Several algorithms are used to explain the component analysis, partial matching, state-space transitions, and tree sequence.

References

1. Uthayakumar, J., Vengattaraman, T., & Dhavachelvan, P. (2018). A survey on data compression techniques: From the perspective of data quality, coding schemes, data type, and applications. *Journal of King Saud University – Computer and Information Sciences, Elsevier*. https://doi.org/10.1016/j.jksuci.2018.05.006.
2. Li, L., Muneeswaran, V., Ramkumar, S., Emayavaramban, G., & Gonzalez, G. R. (2019). Metaheuristic FIR filter with game theory based compression technique – A reliable medical image compression technique for online applications. *Pattern Recognition Letters, Elsevier, 125*, 7–12.
3. Dutta, T. (2015). Medical data compression and transmission in wireless ad hoc networks. *IEEE Sensors Journal, 15*(2), 778.
4. Chen, W., Chen, Z., & Cui, F. (2019). Collaborative and secure transmission of medical data applied to mobile healthcare. *BioMedical Engineering OnLine, 18*(1), 60.
5. Kolo, J. G., Shanmugam, S. A., Lim, D. W. G., Ang, L.-M., & Seng, K. P. (2012). An adaptive lossless data compression scheme for wireless sensor networks. *Hindawi Publishing Corporation Journal of Sensors, 2012*, 539638.
6. Zhu, C., & Xu, H. (2015). Design and implementation of lossless compression encoding for high-speed data acquisition and storage. In *IEEE 12th International Conference on Electronic Measurement & Instruments*. https://doi.org/10.1109/ICEMI.2015.7494273.
7. Sunil Kumar, B. S., Manjunath, A. S., & Christopher, S. (2016). Improved entropy encoding for high efficient video coding standard. *Alexandria Engineering Journal, Elsevier, 57*(1), 1.
8. Hameed, M. E., Ibrahim, M. M., Manap, N. A., & Mohammed, A. A. (2019). A lossless compression and encryption mechanism for remote monitoring of ECG data using Huffman coding and CBC-AES. In *Future generation computer systems*. New York: Elsevier.
9. Venugopal, D., Mohan, S., & Raja, S. (2016). An efficient block based lossless compression of medical images. *Optik, Elsevier, 127*, 754–758.
10. Li, R., Hao, Z., Zhan, W., Zhan, W., & Usman, A. T. (2020). A medical image compression and reconstruction method based on improved principal components analysis. *Acta Microscopica, 29*(1), 521–527.
11. Reetu Hooda, W., & Pan, D. (2020). *Tree based search algorithm for binary image compression*. Huntsville: IEEE Xplore, IEEE.

Chapter 5
Wearable Smart Devices for Remote Healthcare Monitoring to Detect Cardiac Diseases

Ashok Kumar Munnangi, Ramesh Sekaran, Geetha Velliyangiri, Manikandan Ramachandran, and Ambeshwar Kumar

5.1 Remote Healthcare Monitoring

RHCM is expanding in acceptance as both patients and healthcare office workers desire health to be kept track of the exterior of clinical environments. RHCM is also named remote patient observation of the exterior of clinical settings. In other words, RHCM refers to the procedure of utilizing technology to observe patients in non-clinical settings, like inside the house.

5.1.1 Introduction

Relative Humidity Control Monitor (RHCM) has been proposed as a proportion of a prompt interferences and precaution healthcare prototype in which patients are observed over not just at an intermittent doctor call on. Now a days as outpatients using mobile devices and wireless analysis strategies. Contemporary reinforce this model of care, manifesting improvement in patient quality of life, and minimum cost incurred in healthcare. The essential indiction of stalking for chronic conditions was utilized with quick interposition.

A. K. Munnangi · R. Sekaran
Department of Information Technology, Velagapudi Ramakrishna Siddhartha Engineering College, Vijayawada, Andhra Pradesh, India

G. Velliyangiri
Department of Electronics and Communication Engineering, Kongu Engineering College, Perundurai, India

M. Ramachandran (✉) · A. Kumar
School of Computing, SASTRA Deemed University, Thanjavur, India

© The Author(s), under exclusive license to Springer
Nature Switzerland AG 2021
D. J. Hemanth et al. (eds.), *Internet of Medical Things*, Internet of Things,
https://doi.org/10.1007/978-3-030-63937-2_5

5.1.2 Significance and Components of Remote Healthcare Monitoring System

The design of technology making changes to both patients and healthcare executive intensive situations are said to be attended at an earlier stage. Though the costs involved are found to be higher, monitoring of patients in a remote manner is probably to become a chief element of the precautionary health protection in the future. Some of the significance of RHCM systems are listed below:

- Healthier gain to healthcare
- Better quality of life
- Harmony of mind and everyday assertiveness
- Better assistance, schooling, and assessment

5.1.2.1 Remote Healthcare Monitoring Assistances to Patients: Healthier Gain

In a nation where a sizeable escalation in the frequency of insured has made it more cumbersome for certain patients to acquire caregivers, RHCM boosts the potentiality for healthcare professionals to treat considerable patients. The likelihood of more healthcare establishments cuddling remote healthcare monitoring unlocks the door to heightened access for patient care globally.

5.1.2.2 Remote Healthcare Monitoring Assistances to Patients: Better Quality of Life

Moreover, RHCM also has the potential in enhancing the patient's quality of life and patient care. This is due to the fact that RHCM connects healthcare professionals instantly virtually with pertinent patient information. It makes their day-to-day chores efficient and mitigates the probability of expenditure resulting in apparent advantages to patient care.

5.1.2.3 Remote Healthcare Monitoring Assistances to Patients: Harmony of Mind and Everyday Assertiveness

Salient as patient relief and consultation are, the advantages of RHCM go beyond that, providing patients the assurance that someone is looking out for their healthcare and welfare daily. Let us consider a cardiac patient whose heartbeat is being monitored for days, prompting the patient to feel very much scared. Providing a crossover and an association for that patient in the home will create an image that the healthcare group goes along with the patient.

5.1.2.4   Remote Healthcare Monitoring Assistances to Patients: Assistance, Educating, and Assessment

With RHCM assistance provided to essential patients, it also bestows thickened levels of assistance, knowledge with the requirements of remote monitoring, and assessment in an intermittent manner compared with the conventional healthcare model.

5.1.2.5   Components of the Remote Healthcare Monitoring System

The devices utilized in RHCM are analogous to smartphones and tablets. It is constructed to acquire evaluation and link with a particular prerequisite or a healthcare professional for transmission of medical information. Patients would need to employ particular sensors that transfer patient physiological data to healthcare professionals. Figure 5.1 shows a block diagram of the components of the RHCM system.

The healthcare professionals then utilize this patient medical information so that timely assessment of the patient's conditions is made and assistance is said to be given accordingly. A detailed description of the block diagram is presented below.

Input Devices

Devices that stay with the patients besides the sensors that not only provide a systematic transfer of the patient's conditions via information about several factors but also support to the patients by administering certain factors as follows:

- Reminders
- Caution messages
- Appointments with the doctor
- Providing intellect cooperation

Here, the patient enters the data manually, or the sensors automatically detect and input the data into the device. Moreover, input devices may also be a sensor connected onto a patient's garments, mundanely adhered to a patient by one or more steer, or implanted in a watch, cleat, clothing, and so on. Other forms of input devices include:

- Computers
- Android phones
- Landline phones

The above said equipment are utilized for patients to take up data manually or wirelessly through a local storage device. These devices are key elements that make a remote healthcare system different from any other healthcare system.

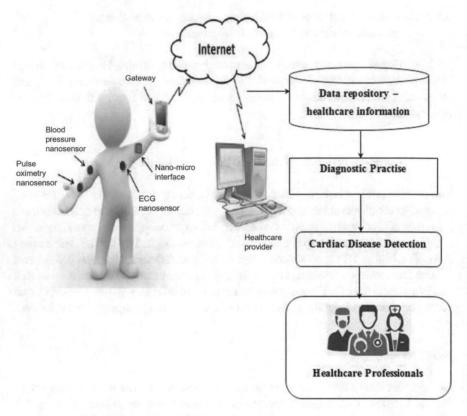

Fig. 5.1 Block diagram of components of remote healthcare

Data Repository

Relevant or pertinent information acquired from the patient has to be stored in a device which can be extracted at any point in the treatment. Hence, it is crucial that all information analogous to the patient are kept in devices like the Universal Serial Bus (USB) or in a repository which can be acquired whenever essential.

Diagnostic Practice

For a specified treatment sequence, it is indispensable to grip the information about the patient's health specifications in a USB or in a repository to be acquired whenever the need arises.

Communication Network

A network is an alternative segment of the solution that specifically assists in establishing RHCM with the patient's input device. Different types of networks are said to be available in the market nowadays, and it is predominant to make certain the potentiality of the input device to connect to the network in concern. This is said to be achieved from:

- Wireless Fidelity (Wi-Fi)
- Local area networks
- Mobile ad hoc networks
- Bluetooth establishments

The provider of the services usually acquires this network to institute patient-doctor connections.

Central Repository

Patient data is stockpiled in remote healthcare systems conserved by healthcare centers and nursing homes. This could be established by one or more of the centralized repositories relevant to a healthcare system. Healthcare software, in turn, utilizes and provide this updated data to healthcare professionals and furnishes the information corresponding to the condition of the patient for further verification, remedy, and therapy.

5.1.3   Infrastructure of IoT

Internet of Things, as a strong area of computer networks, necessitates a conventional framework to provide an aggressive circumstance for medical healthcare professionals to improve their quality of life. Moreover, a comprehensive assessment of the conventional internet frameworks necessitates being conducted to evaluate its potentialities converge to provocations of the internet of things.

The Internet of Things associates a great number of heterogeneous devices via the Internet. It should sustain a ductile layered architecture. Different types of architectures are said to be designed for the IoT [1]. A few examples are as follows:

- Three-layer architecture
- Middleware architecture
- Service-based architecture
- Product-based architecture
- Five-layer architecture

Figure 5.2 shows a sample of a five-layer architecture.

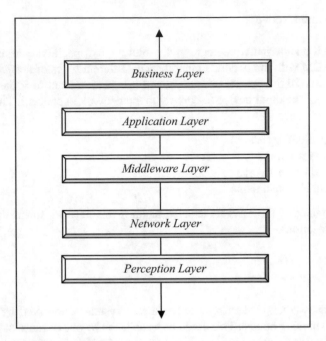

Fig. 5.2 Sample five-level architecture

5.1.3.1 Perception Layer

Initially, the layer starts with the perception of objects in the concerned layer. In this perception layer, the sensor devices acquire and progress the information, computerize the collected information from the patient, and then send the patient-sensitive information to the network layer. The perception layer comprises sensors for acquiring information like location, temperature, weight, humidity, and so on. In other words, passage of enormous data produced from disparate systems and stocked in the Internet is initiated in the perception layer.

5.1.3.2 Network Layer

In the network layer, patient information acquired from the perception layer is dispatched via a secure media to the middleware layer. Here, data are said to be transferred by utilizing technologies, which include but are not limited to:

- Radio-frequency identification (RFID)
- Global System for Mobile Communications (GSM) between devices
- Wi-Fi
- Bluetooth

The sensitive medical data acquired from the abovementioned objects are further stockpiled and refined by utilizing cloud services. A Convolutional Neural Network (CNN) based regular pattern mining model for the discovery of knowledge related to regularity in health conditions. A new convolutional neural network (CNN) learning model was introduced in [2] to identify the correlated health-related factors. A double-layer fully connected CNN structure was applied for categorizing the gathered data.

5.1.3.3 Middleware Layer

The middleware layer splits the service according to the name and address with the sensitive medical information acquired from the medical health professional. This layer in other words permits developers to operate with diversified devices irrespective of the hardware plan of action. Different intentions for the requirement of possesing an interface in IoT. First, as various technologies are prevalent on the IoT, routine calibrations are said to be a very cumbersome process.

Therefore the IoT must be coupled to heterogeneous computers. Second, there is said to exist an insistence for isolating the application layers in disparate realms. To address this, an interface is provided by the middleware. Finally, the layer also conceals irrelevant details and diversification of physical layer mechanisms to untangle the comprehensive procedures, specifically to the patients.

5.1.3.4 Application Layer

This application layer is said to be established upon patient request, entirely depending on the practicable desirability of entities like:

- Blood sugar measurements
- Blood pressure measurements

The significance of this application layer owes to the conception that it can bestow high-quality intelligent assistance to converge several patient prerequisites.

5.1.3.5 Business Layer

Finally, the business layer controls all system pursuits and assistance. The culpability of this business layer is to construct a business framework, graphical structures, and flow patterns based on the medical data acquired from the patients via the application layer.

5.1.4 Remote Healthcare Monitoring Architecture for Cardiac Disease Detection

The RHCM architecture for cardiac disease detection [3] involves a three-level model. They are:

- Wearable sensor
- Android/Apple device
- Web gateway

The three-level architectures are shown in Fig. 5.3.

5.1.4.1 First Level

The first level of cardiac disease RHCM involves the interface of the patient that comprises several multiple wearable sensors. These are utilized in obtaining the medical information about each patient collected at a different time interval. This first level transfers the data in a real-time manner in a wireless manner from wearable sensors that are implanted in the patient's body parts to the second level via Bluetooth.

5.1.4.2 Second Level

The second level comprises either an Android phone or an Apple phone. With these devices, the patient's information is extracted or acquired from the wearable sensors. Android phones or Apple devices have the potential to interface with the web-server using General Packet Radio Services (GPRS) or Wi-Fi networks.

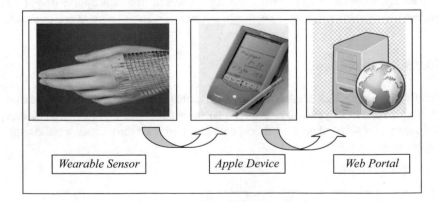

Fig. 5.3 Cardiac disease remote healthcare monitoring

5.1.4.3   Third Level

Finally, in the third level, the web server obtains the data or information from the database and hands over the data to the online database. The web server is a podium that obtains the data of patients from all over the globe with wearable sensors and portrays them on the web interface. The information here includes patient location information and his confidential information for recognition. Some of the information that can be extracted are the heart rate, blood pressure level, blood sugar level, and so on.

5.2   Communication Technologies for Remote Healthcare Monitoring to Detect Cardiac Diseases

The performance of a middleware that extends a forever distinguished programming pattern and dissociates the originator from low-level concerns, specifically the various communication technologies utilized. It is very predominant and critical to the specific extent for RHCM to detect cardiac diseases.

To keep away from probable interventions with the subsystem and also with certain other extensive communication, the IEEE 802.11 is utilized to detect cardiac diseases. This is the most remote of the channels utilized to transmit location beacons. Different communication technologies are therefore in use for remote healthcare monitories, specifically to detect cardiac diseases. To name a few are:

- Bluetooth (BLE)
- Wi-Fi
- Zigbee
- Long-range devices (LoRa)

In recent years, to impart uniform cardiac health data wirelessly to the cloud environment, LoRa is utilized. This device is a novel, exclusive, and spread-spectrum modulation strategy that permits sending the measured data from the patient at exceedingly minimum data rates to immensely extensive ranges.

5.2.1   Outline of the Remote Health Monitoring System

Automated monitoring has been shown to be effective in helping the elderly or senior citizens in the management of chronic health concerns. In an early stage with healthcare provider interference is inherently outshine. The patients who may wait for a longer period become earnestly ill before going on the lookout for treatment.

To integrated into the administration of chronic ailments, RHCM has the prospective inefficiently enhancing the patient quality of life and hence its popularity has also been heightened. In RHCM, a monitoring device needs a sensor that can assess specified data about physiological aspects. This information is passed on to both the patient and healthcare executives in a wireless manner. With the inception of sophisticated devices, RHCM enables patients to keep an eye on their health to manage chronic conditions. Two types of remote health monitoring systems are there in existence. They are shown in Fig. 5.4.

As depicted in the above figure, the elaborate description of the communication technologies, short-range, and long-range are provided below.

5.2.2 Short Range Communication Technologies

The selection of the communication technology to be chosen based on the monitoring necessitates for cardiac diseases is considered to be a paramount issue when constructing for specific disease detection. The selection of communication technology certainly is depended on the monitoring prerequisites and the configuration's features. Amongst several communication technologies in the market, one of the few for cardiac detection is ZigBee [4].

- The Zigbee protocol was designed by the Zigbee Alliance. To start with, it was formulated as a standard for home automation. In the later years, it was publicized, and its practicability has been broadened to reshape it to several other applications also.
- Zigbee has the potential to inspect radio channels to record the information involving cardiac detection with the slightest intervention, which is then chosen and utilized by the entire device involved in the Zigbee network.
- Zigbee permits an utmost transmission speed of 250 Kbps connecting devices with a detachment of covering 50 m. Zigbee at present accounts for one of the

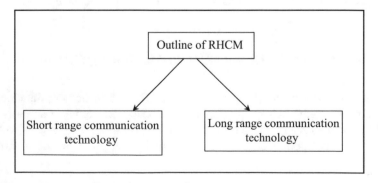

Fig. 5.4 Outline of communication technologies

most comprehensively utilized protocols for cardiac remote healthcare monitoring.

5.2.3 Long-Term Monitoring Systems

Long-term monitoring systems or long-range technologies in comparison with the conventional communication technologies permit kilometer-wide wireless communications. In long-term health monitoring system, an ECG remote monitoring system was introduced in [5] which offers a high-quality ECG signal.

This involves a star topology with radio frequency ranging between 860 MHz and 1020 MHz.

- The data range of long-range monitoring systems lies in between 290 bps and 50 Kbps with an overall range of 15 Km.

Therefore, the LoRa involves a high communication range with a minimum data rate forming the best selection for remote healthcare systems monitoring cardiac diseases. This is because with this range of communication, it ensures the absolute interface between the wearable sensors measuring the cardiac disease data.

5.2.4 Challenges in Remote Healthcare Monitoring

RHCM has the prospect to metamorphose patient care entirely. If executed comprehensively, it will reorganize the circumstance and empower healthcare assistances to supervise their resources much efficiently. Some of the challenges [6] faced are:

- Patient data reliability
- Patient data precision
- Real-time access to patient data
- Network connection and compatibility

5.2.4.1 Patient Data Reliability

The medical data about cardiac disease communicated over any RHCM plan of action to be reliable enough to converge the levels essential of healthcare. It will require vigorous data management executions, comprehensible frontiers of possession, and rigid reliability standards. The challenges are no fewer critical for clinics and sophisticated hospitals that peril a third party that could be muddle along putting their patients' reliability and seclusion at menace.

5.2.4.2 Patient Data Precision

Controversially the most complicated challenge facing RHCM of cardiac disease concerns patient data precision. Medical professionals will be anticipated to recognize and handle patients from the perspective of the data concerned. This is because of the reason that the medical professionals even have to handle or treat a chronic patient with these highest probable accurate data.

5.2.4.3 Real-Time Access to Patient Data

The relocation of information necessitated for RHCM to work is possibly a lengthy and intricate procedure and comprises a voluminous transfer. Initially, data must be acquired via the patient's wearable sensors. If the data being acquired is on a mobile network, then patient data has to traverse via the network provider's framework via voluminous data centers. Upon the occurrence of interruption in any one of the hops, the patient's sensitive and intensive data are said to be hindered in arriving at its stopping place.

5.2.4.4 Network Connection and Compatibility

The victory of RHCM hangs laboriously on the connection and compatibility of the network and the wearable sensors. This is specifically true when interferences could containment the diagnosis, in the case when necessitates continuous monitoring of heart conditions.

5.3 Remote Healthcare Monitoring to Detect Cardiac Disease

In RHCM, the wearable sensors involved had substantially minimized the load of patients and distributed an excessive life care style with minimum menace. Beyond the patient associate, the mechanics manifesto is being strengthened in every predicament and deciphers all patient medical-associated issues remotely. Some of them are listed below.

- Boosting chronic circumstance administration
- Narrowed nascent circumstances and readmissions
- Minimized load on healthcare systems
- Better patient consequences
- Improved quality of life

From the above-said factors, the objective of RHCM is to relay messages in a prompt manner and also early detection of a critical worsening condition.

5.3.1 Early Heart Disease Prediction

One of the well-known diseases that affect several people both during their mid-age and in the senior age is heart disease. In either case, it may also result in a higher mortality rate. Also, heart disease is found to be more common in men than in women. Even it is said that around 17 million people die due to cardiovascular disease (CVD) [7] globally every year and it is found to be highly identified in Asia. For research on heart disease, the Cleveland Heart Disease Database (CHDD) is used as the de facto database. Some of the risk factors involved are:

• Age
• Sex
• Smoking habits
• Hereditary
• Cholesterol level
• Poor diet practices
• Obesity
• High alcohol usages
• High blood pressure
• High diabetes

In addition to the above forces identified, certain other factors also play a major role:

• Eating practices
• Irregular exercise or absence of physical activity
• Obesity

Moreover, different types of heart diseases are said to be prevalent all around the globe. To name a few, the following are samples:

• Coronary heart disease
• Congestive heart failure
• Congenital heart disease
• Myocarditis

Even though above factors are said to play a major role in detecting heart disease, it is a cumbersome process to evaluate the odds of heart disease manually. However, RHCM is utilized to predict heart disease.

5.3.2 Remote Patient Monitoring in Heart Disease Patients

RHCM applications permit minimizing the risk concerning heart failure in patients susceptible to this state with the aid of cardiac resynchronization wearable sensors that can transmit patient cardiac health information to a centralized database. This information can be regularly monitored, and patterns can also be obtained. Therefore assisting patients to maintain stable health, enhancing their quality of life, and reducing the overall fatality rate are important. Figure 5.5 shows a schematic view of remote patient monitoring (RPM) in heart disease patients.

Due to a vibrant way of life, several people are agonizing from heart diseases. To minimize the risk of heart failure, technology has made several advancements including cardiac resynchronization treatment and the pacemaker. This in turn refers to that the RHCM enhances the patient quality of life, reduces the fatality rate, and minimizes the period of stop off in hospitals.

5.3.3 Cardiovascular Monitoring System and Remote Monitoring

In the past few years, medical wearable sensor makers had analogously manageable chores to conduct specifically intensified on monitoring hospitalized patients and fabricating tools. Tools such as pacemakers have helped eliminate or prevent problems in cardiac patients' health enabling them to live a normal life.

However, nowadays, with the maturity of digital healthcare, even manufacturers and medical professionals are under vigorous pressure to do something extra—collecting information and compiling for medical analysis in a real-time environment. To be specific, they have been asked to identify new methods of timely patient data

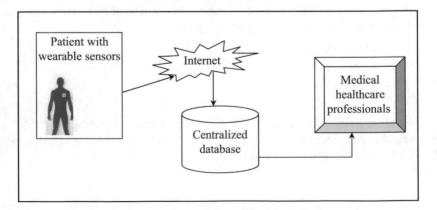

Fig. 5.5 Schematic view of heart disease remote patient monitoring

collection, so medical professionals can acknowledge swiftly pertinent changes in the patient's status.

Remote cardiac monitoring, in particular, has emerged as a target for device makers who realize that their products can close gaps in the existing toolset and offer lifesaving benefits to a fragile population. Ultimately, RPM helps providers offer decentralized services that are gradually replacing many of those delivered inside hospital walls. With Medicare paying for remote monitoring, providers have every reason to get on board with RPM.

Continuously monitoring the heart assists medical health professionals in acquiring the information to evaluate the heart's health state. In specific time, devices involved in cardiovascular are utilized in diagnosing the heart condition, or else they are also utilized in assisting a heart ailment. A novel two-step predictive framework was introduced in [8] for high-risk heart patients. The framework is important for other biomedical signals, namely, photoplethysmography (PPG), pulse oximeter (Pleth), and electroencephalogram (EEG).

5.3.4 Remote Cardiac Monitoring

For at-risk patients, detected with cardiac deformities resulting in heart symptomatic circumstances, medical health professionals utilize cardiac monitors to identify the severity. These cardiac monitors continuously monitor heart activity and send intensive data to electronic health records, therefore, empowering more comprehensive problem-solving.

With the help of these devices, even perforations are not even left due to the reason that it does not have to be separated for charging on a day to day base. This recognize unusual indications for consistently keeping an eye on high-risk patients. The processes involved are as follows:

- Through remote cardiac monitoring, critical information obtained from the patient's wearable sensors are said to be communicated to a medical healthcare professional directly.
- With this consistent and continuous monitoring, medical healthcare professionals are permitted to assess the patient's heart activity without the presence of the patient physically.
- The main reason for effective monitoring is due to the small device nature acquiring and sharing the most prerequisite information is said to be done in a significant manner. This is said to be achieved whenever the cardiac monitoring device coincides with the transmitter and shares the collected data.
- The transmitter then transmits the vital information with the medical healthcare professional via a secure web server that is said to be accessed only by the patient's medical healthcare professional team.

Some of the benefits of remote cardiac monitoring to mention a few are:

- Lesser in propria persona is required
- Economizing patients time and money
- Minimization in hospitalizations
- Curtailing time consumed in recognizing clinical events
- Higher probability of survival rate
- Fewer follow-up visits
- Found to be extraordinarily salient in this era of COVID-19

5.4 Smart Healthcare Future Challenges and Barriers Relevant to Cardiac Diseases

The future of organizational alimentation in cardiac healthcare necessitates capable heads to intent for any prospective issues. In the next 5 to 10 years, medical healthcare professionals should anticipate experiencing an abundance of issues. The following just to name a few are examples:

- Administrative and strategy changes
- Developments in medicine and technology
- Issues about sponsoring and education

Medical healthcare professionals should understand the fusion of these ultimatums. Some of them are:

- Time and money for cardiac disease medical research
- Providing state-of-the-art facilities
- Utilization of modernized equipment and wearable sensors
- Educating both medical healthcare professionals and patients

Connected health utilizes smart and mobile applications with wireless devices like Bluetooth, Wireless Fidelity, and long-range devices so patients are connected with their medical healthcare professionals without having to visit them in person routinely. Connected health has progressed into smart healthcare wherein the traditional smartphones are utilized with wearable sensors to monitor blood pressure, glucometers, and smartwatches to validate patient monitoring uninterruptedly even when patients are in their homes. Smart cardiac healthcare is anticipated to keep expenses related to hospitals low and obtain treatment in a timely and precise manner.

5.4.1 Authentication and Authorization Issues

Smart healthcare devices can give rise to patient security and privacy. A novel Integrated Circuit Metric (ICMetric) technology was presented in [9] which involves providing a security key, an authentication scheme applied to authenticate separate devices. Some of the issues faced by both the medical healthcare professionals and patients are listed below:

- Unauthorized access to smart healthcare devices could generate a significant risk to patients' health as well as to their sensitive medical personal information.
- Associated instruments together with healthcare and mobile wearable sensors acquire, combine, evaluate, and transmit sensitive medical information to the cloud computing environment. Here the device layer is said to be highly susceptible to spoofing attacks, cloning, and so on.
- Denial of service (DoS) attacks can influence medical healthcare systems and even influence patient protection to a greater extent.
- The abstain discernment of prospectively security menaces residues a provocation due to the frequency and complication of transpiring software and hardware susceptibilities.
- The insufficiency of security grades of these smart healthcare devices in addition to the potentiality of strong search engines makes these wearable sensors highly susceptible to all types of attacks [9].
- In recent years, several wireless networking mechanisms have also been designed in the smart healthcare framework, like Wireless Fidelity, Bluetooth, and so on that are being utilized in providing long connectivity to several wearable sensors. Security mechanisms of these wireless devices against Sybil attacks and worm whole attacks must be imposed.
- A centralized repository of information about the personal history of the patient, family background, and medical data has to be preserved and protected from both hackers and malicious software to provide proper security and privacy mechanisms.

Moreover, certain risks involved are the geographical information of the patient with a purchase made from the medical practitioners or pharmaceutical companies that may bestow a description of a patient's health status.

5.4.2   Risk Analysis

As far as smart healthcare is concerned, risks to patients, smart healthcare workers, and medical healthcare professionals are widespread in healthcare also. Hence, it is a prerequisite for any hospital to possess a certified medical healthcare risk manager to evaluate, expand, execute, and observe a risk management plan of actions to reduce the disclosure. The role of smart medical healthcare risk managers [10] includes the following:

- Smart medical healthcare risk managers are instructed to operate several issues in multifaceted environments.
- Smart medical healthcare risk managers operate in the following areas:

    Financing, insurance, and claim analysis

Event, accident, and insurance management
Clinical settings
Emergency activities

- Risk managers work in both static and dynamic manner either to safeguard incidents or to reduce destructions.
- Some of the challenges faced by administrators are

Safety hazards for the patient
Mandatory rules and regulations
Probable medical error
Prevailing and succeeding policy

Hence, probable risks have to be measured and analyzed frequently. Also, their side effects have to be analyzed. Based on the risk factor analysis, smart medical healthcare managers should design, develop, and implement the plans accordingly.

5.4.3   Standardization and Validation Plans

The confronts suffered by standardization and validation plans about the prospective of smart healthcare wearable sensors to provide cost-efficient, improved experience quality and different smart healthcare services are evaluated by the complicatedness of upsurging patients.

Smart healthcare environments are exceedingly complicated and stimulating to manage as they are essential to managing with a diversity of patient circumstances under differing occurrences with a plethora of resource limitations. Several statutory bodies have designed a greater number of subsidies to bestow a robust plan of action to counter the ultimatum of providing smart cardiac healthcare solutions.

To address most of the healthcare standardization and validation issues, utilization of specific technology can assist in attaining the objective of comprehensive health protection. With mushroom growth in both technology and wearable sensors, gaining medical assistance to all sectors of the public is becoming a reality. With the most sophisticated standards, telemedicine has made easy accessibility and availability in the urban and rural sectors of the country. Also, telemedicine has been used in rural sectors where the inadequacy of medical healthcare providers and assistance provided directly due to lack of accessibility has been minimized.

5.4.4   Network Availability and Connectivity

As far as smart healthcare is concerned, it is undergoing a fast amendment from conventional hospital and professional concentric pattern to a dispersed patient-centric manner. Among several techniques involved, network availability and connectivity fire this fast metamorphosis of healthcare in a vertical fashion.

In the present scenario, healthcare broadly utilizes the prevailing 4th generation network and is constantly maturing to reconcile the requirements of future intelligent healthcare applications. With the expansion of the smart healthcare market connecting to the network, it will generate enormous data different in both size and formats also. This, in turn, will lead to complicated network traffic requirements in terms of bandwidth, data rate, and latency.

The developed state, the smart healthcare framework, also the connectivity has to be provided for a higher number of wearable sensors or devices. This in turn would necessitate the requirement to design feasible machine-type communication. However, prevailing communication technologies can compete with the existing data and devices in the network placed by different smart healthcare providers and devices in the market. Hence, the upcoming 5G network is anticipated to underpin smart healthcare applications that aid in ultralow latency, high bandwidth, ultrahigh reliability, high density, and high energy efficiency.

5.5   Conclusion

In this chapter, an overview of RHCM carried out by previous researches is presented. Various communication technologies are used for RHCM to discover cardiac diseases. RHCM is used for identifying the cardiac disease to reduce the heart failure risk. Future challenges are analyzed for authentication and authorization issues, risk analysis, standardization and validation plans, and network availability and connectivity.

References

1. Alansari, Z., Anuar, N. B., Kamsin, A., Belgaum, M. R., Alsh, J., Soomro, S., & Miraz, M. H. (2018). Internet of Things: Infrastructure, architecture, security and privacy. In *IEEE International Conference on Computing, Electronics & Communications Engineering* (pp. 150–155). Solan: Jaypee University of Information Technology.
2. Ismail, W. N., Hassan, M. M., Alsalamah, H. A., & Fortin, G. (2019). CNN-based health model for regular health factors analysis in internet-of-medical things environment. *IEEE Access, 8,* 52541–52548.
3. Kakria, P., Tripathi, N. K., & Kitipawang, P. (2015). A real-time health monitoring system for remote cardiac patients using smartphone and wearable sensors. *Hindawi Publishing Corporation, International Journal of Telemedicine and Applications, 11*(1), 11.

4. Alonso, L., Barbarán, J., Chen, J., Díaz, M., Llopis, L., & Rubio, B. (2018). Middleware and communication technologies for structural health monitoring of critical infrastructures: A survey. *Computer Standards & Interfaces, Elsevier, 56*, 83–100.
5. Spanò, E., Di Pascoli, S., & Iannaccon, G. (2016). Low-power wearable ECG monitoring system for multiple-patient remote monitoring. *IEEE Sensors Journal, 16*(13), 5452–5462.
6. Sagahyroon, A. (2017). Remote patients monitoring: Challenges. In *2017 IEEE 7th Annual Computing and Communication Workshop and Conference (CCWC)* (pp. 1–4). Las Vegas: CCWC.
7. Tian, S., Yang, W., Le Grange, J. M., Wang, P., Huang, W., & Ye, Z. (2019). Smart healthcare: Making medical care more intelligent. *Global Health Journal, Elsevier, 3*(3), 62–65.
8. Chen, J., Valehi, A., & Razi, A. (2019). Smart heart monitoring: Early prediction of heart problems through predictive analysis of ECG signals. *IEEE Access, 7*, 120831–120839.
9. Tahir, H., Tahir, R., & McDonald-Maie, K. (2018). On the security of consumer wearable devices in the Internet of Things. *PLOS One, 13*, 1–21.
10. Ahad, A., Tahir, M., & Yau, K.-L. A. (2019). 5G-based smart healthcare network: Architecture, taxonomy, challenges, and future research directions. *IEEE Access, 7*, 100747–100762.

Chapter 6
Smart Assistance of Elderly Individuals in Emergency Situations at Home

Anu Radha Reddy, G. S. Pradeep Ghantasala, Rizwan Patan,
R. Manikandan, and Suresh Kallam

6.1 Introduction

The availability of wellness services for older individuals is an urgent need at present. Many elderly people experience isolation and emotional distress as a consequence of living alone or without a strong support system. To help with these issues, we have developed an Internet of Things (IoT)-based device that integrates wearable equipment, biosensors, and a wireless antenna to provide assistance to and monitor the physical health of older individuals. In this system, the data gathered from different wearable pieces of equipment are maintained in a central database, thus allowing caregivers and healthcare providers to access information in an emergency situation. The method is convenient, effective, and decreases the cost of health care while protecting and monitoring elderly people in their day-to-day routines.

A. R. Reddy
Department of Computer Science and Engineering, Malla Reddy Institute of Technology and Science, Hyderabad, Telangana, India

G. S. P. Ghantasala
Department of Computer Science and Engineering, Chitkara University institute of engineering and technology, Punjab, India

R. Patan (✉)
Department of Computer Science and Engineering, Velagapudi Ramakrishna Siddhartha Engineering College, Vijayawada, India

R. Manikandan
School of Computing, SASTRA Deemed University, Thanjavur, TN, India

S. Kallam
Department of Computer Science and Engineering, Sree Vidyanikethan Engineering College (Autonomous), Tirupati, India

D. J. Hemanth et al. (eds.), *Internet of Medical Things*, Internet of Things,
https://doi.org/10.1007/978-3-030-63937-2_6

A stroke is an attack or infarction of the brain triggered by a blood supply disruption to the region. Although the number of deaths linked to stroke has decreased in recent years, the rate and prevalence of stroke is increasing, remaining as the principal cause of death in Korea. The population affected by stroke and the global population are aging. On average, in Korea, one person experiences a stroke every 5 min. A patient who has an acute stroke will benefit from having someone nearby to identify the warning signs; in certain stroke cases, the patient will fail to seek assistance on his or her own accord. However, the probability of recovering from acute stroke greatly improves if the individual obtains medical attention right away.

The most important area of use for wearable electronics is wireless health tracking. Digital computing and remote health monitoring can be combined for "smart" healthcare tracking using the IoT. One of the major focuses of biomedical technologies is the management of healthcare and emergency response for older individuals. The primary aim of our research was to address environmental-related living concerns for effective identification and alert triggers for the onset of stroke, which permit prompt delivery of medical care to reduce the long-term effects of such attacks.

IoT Assistance Currently, the IoT connects a range of daily tools and devices, such as antennas, actuators, machines, workstations, and mobile phones, which can integrate with many types of equipment and exchange information [1], hence providing an efficient solution to incorporate e-health and assistance for the aged.

HABITAT (House-Aid Based on the Internet of Things for the Autonomy of Everyone) [2] integrates IoT-based tools in the home, such as radiofrequency identification (RFID), wearable electronics, artificial intelligence (AI), and wireless sensor networks (WSN) [3–5]. HABITAT has the potential to support and protect older individuals in their residences, allowing them to safely remain in their homes and perform daily activities.

HABITAT's main objective is to investigate an IoT-supported solution to provide an assistive, reconfigurable living room using a standard interface. Thus, a design featuring interoperability and reconfigurability using different tools [3] has been employed. A focus is on scalability and customizations to monitor the day-to-day activities of individuals who require special attention in their home because of advanced age or illness [4]. Communication skills are handled by AI methodologies.

In addition to improvements in hospice equipment and pharmaceutical treatments, a novel approach in IoT technology is the ability to provide practical improvements for the aged population. For example, safety functions are available on smartphones and other equipment [5]. These opportunities can be improved with further machine-to-machine connections [6]. However, healthcare services are facing challenges because of the rapid growth of the elderly population [7]. The IoT allows an individual's health data to be monitored using equipment with RFID credentials to observe patients remotely.

An effective monitoring program is unobtrusive, being able to identify the target parameters without affecting the user's normal behavior. To this end, positioning and body motility are the most relevant parameters for monitoring a user's behavior. Recent advancements in wearable and mobile technology are providing major improvements in this direction. Wearable devices are fitted with various sensors, such as gyroscopes, accelerometers, and global positioning system (GPS) to detect location and motility [8].

H2U monitors daily fitness for the aged people based on their health conditions it will suggest them and guide them to perform various physical exercises. The development is intended to authenticate the effectiveness of assorted skills by monitoring the everyday activities of elderly individuals using tools such as wearable antennas and actuators. Research is being conducted to demonstrate the effectiveness of the scheme. The systems operate with high strength and are extremely consistent for senior citizens (Fig. 6.1).

Contribution

The system will recognize changes to vital signs, which can trigger warnings in emergencies. The wearable tool can be incorporated into an inconspicuous and comfortable bracelet, providing an ideal solution for any age group to use at home. A web application gathers all data collected and sent to the server by the bracelet. It can remotely alert a caregiver or healthcare provider to any emergency. The stored data may then be analyzed to help the healthcare staff track a patient's progression. The implemented IoT structural design can be integrated with existing technology to perform other protocols.

This initiative leverages technological advances and the protection of information for older people who require extensive care. It enables the generation of reports such as health metrics, treatment responses, and more efficient therapies for certain diseases. Preliminary research indicates that 65% of the users engaged in exciting events. The system design can monitor activity, movement speed, sleep, and collect local data through sensors for caregiver points of interest, representing frequently

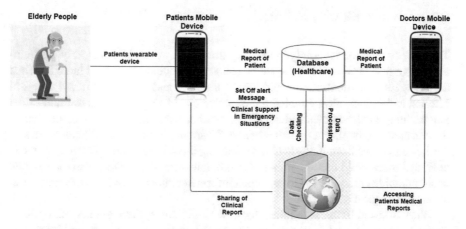

Fig. 6.1 Design of the H2U wellness system

visited locations. The program has faced challenges in the development of a reliable application that collects nearby data. When safely deployed and tested with fewer user nodes, Apple devices stay behind a problem for flare data revealing with Android devices.

Motivation

As discussed, this research's primary objective was to build an IoT-based assistance and monitoring program for elderly users. In the event of a stroke, the device will detect the condition and issue alerts, which will allow for prompt medical treatment to reduce the long-term effects of the stroke. Using IoT, wearable healthcare apps efficiently collect and exchange information between the patient and medical staff in a database network, resulting in more reliable and quicker identification of emergency situations. IoT has great potential in the areas of healthcare and recovery, where the technology can be used to expand access to treatment, enhance urgent care, and improve the quality of care.

This effort uses an ambient assisted living (AAL) architecture that was designed to support seniors in their day-to-day lives. Two essential features are included in the proposed system: 1) it operates in both outdoor and indoor environments and 2) it provides continuous monitoring of health status via the environmental parameters of this information. The system may trigger immediate incidents (e.g., warning family members or medical personnel) when specific hazards emerge. The system also gathers sensor data from various tools transparently and can send it to the remote log server to trigger relevant warnings or to produce announcements. Continuous communication occurs between the AAL program and home automation systems to make it easier to acclimatize to the indoor climate. We aimed to produce a low-cost solution to assist elederly individuals in their activities of daily living, such as cooking and personal hygiene.

6.2   IoT for Elderly People

The aged population is rapidly increasing around the world. When elderly individuals remain in their homes, IoT systems can keep an eye on them and indicate when they may require attention to avoid injury or risk of sudden death [5]. Advances in Internet technology have enabled the effective use of multiple sensors fitted into a single, integrated healthcare device that can be worn on the body or carried to remote areas to monitor a patient's health. Wearable, internet-based healthcare surveillance systems have made health monitoring incredibly uncomplicated, transferable, and more rapid based on ease of communication [6–9]. They also incorporate programmable emergency alarms, which inform healthcare providers of urgent situations that require assistance.

When assisted living is required for older individuals, there are several options, such as day care for adults, home care, nursing homes, hospice care, and long-term care [10]. Although these choices address wellness, medical, social, and day-to-day living needs, they are associated with a loss of autonomy. Most elders enjoy staying at home, where they feel more comfortable than in nursing homes or adult treatment centers. Therefore, older adults may choose to live independently, despite the risk of injury or accidental death [11].

The new IoT-based health monitoring systems allow for effortless, convenient, and quick message-based responses [12]. In healthcare systems [13], IoT concepts allow for customized, proactive, and focused treatment in which patients monitor and manage their health and share accountability with healthcare providers. The care of elderly individuals and people with disabilities, known as AAL, is especially gaining interest because of the rapidly aging population. AAL skills allow individuals to continue with their daily activities and live independently.

A variety of antennas, actuators, and artificial intelligence methods have been explored for use in these systems. At the same time, the discovery of vital circumstances through composite occurrence dispensation originated. The discovery of the subject's plunge and support to employment motion using wearable tools as an alternative has been discussed elsewhere [14]. Additional learning on AAL addressed the modelling of individual and ecological dynamics, which was achieved through process removal methods. In contrast, machine learning methods can recognize users' actions with antenna-originated data.

Modern advancements in portable and wearable antennas have helped in the realization of the vision of AAL. Advanced mobile equipment features a variety of antenna types, including gyroscopes, accelerometers, and global positioning system, to recognize the user's movements and physical well-being [15]. Investigators have developed inconspicuous antennas in the form of small, wearable tools to monitor health indicators. For example, blood pressure, blood glucose, and heart rate can be calculated by wearable sensors. Several other measures, such as electroencephalography, still require larger antennas such as electrodes.

An IoT-enabled health monitoring system has enormous benefits over traditional systems. Patients can be effortlessly equipped with a smart fitness system, thus permitting regular monitoring. Thus, the system is both essential and beneficial to older individuals who require constant supervision. As regular supervising of patients is not feasible by practitioners with a chronometer, IoT facilitates H2U health concerns to be distantly monitored as particular information can be transmitted directly through the Internet in real time [16]. A wireless antenna located in the home is necessary, which can identify emergency situations and hazardous circumstances based on the information it receives. Furthermore, the patient also can be supervised outside the home. It facilitates crisis alerts, which are received by an attendant [17], analyzed, then discussed with a physician as needed.

Security concerns exist with IoT health equipment. Hackers may target the IoT server or the equipment itself. Thus, there are confidentiality issues related to the IoT. The confidentiality of a patient's sensitive health information should be protected during transmission from sensors to portable gadgets and then to a regional server.

The interior contains Bluetooth (BLE) beacons and scanners tested in the Multimedia University's digital home laboratory, which has five rooms: lobby, bedroom, living room, kitchen, and office. The BLE scanner continuously searches for data to upload to the beacon, along with its Received Signal Strength Indicator (RSSI) value to a Messaging Transfer Protocol (MQTT) file. The MQTT server synchronizes the process to ensure that the scanners read the RSSI of the beacon at the same time (Fig. 6.2).

A helpful scheme for health monitoring devices is a descend recognition scheme, which can improve the protection of older individuals. A conventional descend recognition system [18] is a precise tool worn to monitor for urgent situations associated with falls. The device has sensors such as an accelerometer and a gyroscope, which can detect a fall incident. Many investigators [19] have agreed that a smartphone is a suitable alternative to specialized devices. It can be positioned on the waist to detect body movements in the following five patterns: straight-up bustle, position behind, session or immobile position, straight movement, and diminishing. If a fall movement is suspected, then an alert is sent to preselected individuals.

AAL [20] is a policy focused on the advancement of technology and programs to improve the living conditions of elderly individuals and those with chronic illness or disability. AAL applications can provide frameworks for emergency response,

Fig. 6.2 The locality of the wearable BLE beacon

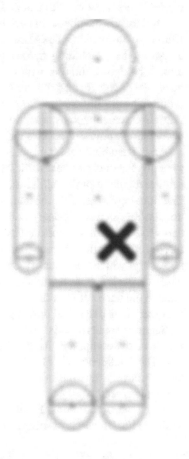

fall detection, and video supervision. Other AAL features have been developed to provide regular help, monitor daily activities, create alerts, and allow older adults to work together with their family unit and healthcare providers.

Substantial research efforts in IoT-driven healthcare services have been made in recent years. Many have their origins in networks of integrated sensors. Suntiamorntut et al. investigated a low-cost, integrated monitoring system that identifies and tracks patients inside nursing facilities. A current trend in IoT is to move outdated modes of operation to structured IP-based networks. For example, an IoT-based smart hospital system (SHS) [20] offers innovative services for automated surveillance and tracking of patients, staff, and biomedical equipment at hospitals and nursing centers.

Arboreta [21] developed a homestay healthcare system for elderly patients with chronic heart and respiratory illnesses. The device features a single node with a wireless sensor that monitors temperature, heart rate, saturation of oxygen, and electrocardiography. The Care Store program is a groundbreaking open-source platform to streamline healthcare device promotion and configuration. A component of the Care Store project, the Common Recognition and Identification Platform, offers sensor-based assistance for seamless consumer and safety system recognition.

AAL has a few smart wearable devices in the marketplace. The Body Guardian Heart is a monitor that collects itinerant cardiac telemetry with cardiac event monitoring (CEM) [22] from patients. Data about the patient are identified passively with delivered to the examiner center, wirelessly through a smartphone. A treatment center can enter its patients' data and review notifications on a web portal. Wellness is a system that integrates sensors, mobile announcements, and home automation for an independent living alternative that is both safe and cost-effective. Family members or caregivers can be notified of unexpected changes in habits, which might indicate an urgent situation, using in-house sensor-based real-time information. Smartwatches may also be helpful for older adults. Although they do not provide monitoring of specific health parameters, they do have accelerometers to detect falls and vocalizations to provide health warnings [23].

We created We-Care, an IoT-ready wireless device for elderly individuals. It can track and gather essential health data from patients, transmit it to healthcare professionals and designated caregivers, and assist in independent living. The information is gathered by a wristband. A concierge is sent tracking coordinates within emergency alerts for situations such as falls and critical changes in vital signs. The system was built with low-power and low-cost criteria, making it ideal for everyone to use in their homes. Our solution can be differentiated by the benefit of its straightforward dealings and amalgamation amid networks allowed to IoT; it contributes a stumpy expenditure device that can be used accessible with all while making provisions for other related systems with the most vital services (Fig. 6.3).

The general structural design of the We-Care system consists of three major components: (1) the We-Watch wristband, (2) the panel of We-Care services, and (3) the background services. The We-Watch is composed of a discreet wristband for monitoring and collecting data from sensors. It securely sends data to the We-Care board responsible for running the web services, while also interfacing with the

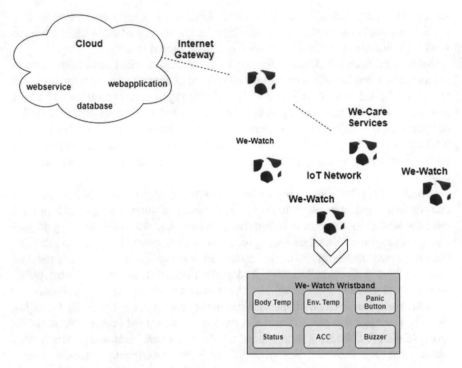

Fig. 6.3 Structural design of We-Care services

cloud when there is an Internet gateway. A standard IoT stack is implemented on the wireless network, and all wristbands use IPv6 underpinned by 6LoWPAN [25]. The We-Care board is responsible for receiving all data from the We-Watch wristbands and running every system service.

In an emergency situation, it activates a distress signal to the concierge and provides a speedy response to any possible problem. Poor connections in the internet might various problems to the millions of users sometimes we might be disconnected from the network because of low or poor signal strength. This board makes the device widely available via an open portal when connected to the internet, whenever a service or apps are available online.

Indoor and Outdoor Localization A GPS receiver and GPRS module, which is mounted on a wearable device, allows the user to be located outside. The GPS module sends details on the user's location to the GPRS module, which is then sent to the Enterprise Service Bus (ESB). When the user is indoors, BLE technology can determine his or her position. One or more inexpensive BLE-equipped devices, called piBeacons [26], can be integrated within all indoor areas of interest (e.g., the user's home).

The piBeacons user's nearest geographical coordinates are sent to the user's wearable computer through a BLE connection. Without ESB, its hard to tackle the

GPRS if it is not associated with IoT. Using BLE technology reduces power utilization, which increases the wearable device's battery life. To provide additional energy savings, the BLE module is deactivated in outdoor locations, whereas the GPS module is turned off while the user is indoors [27].

A detailed categorization of the relationship between RSSI and distance using two BLE tools under controlled conditions is recorded. In testing the best model, the researchers used two inertial magnetic tests units [28], MIMUs, and a laptop for calculating and computing indoor distances. Wearable BLE-based devices were demonstrated to be capable of providing interdistance estimates with a percentage error on average of 26.7% (0.5 m) in a 65-m^2 room.

A novel solution to the indoor localization environment uses an inertial harness measuring units and motion controls (PIR) [2]. The distributed PIR sensors provide location information, while the IMU sensor collects the motion data for observation of body activity. The IMU sensor is used on the thigh, with a vertically aligned axis on the hip. The IMU sensor is used for identification of human behavior such as lying, standing, and sitting. This research demonstrated an indoor positioning and motion tracking strategy that features quick and effective circuit architecture for indoor location accuracy. However, the system is limited to tracking one person.

A key goal of the City4Age project is the collection of sensitive data, which is accomplished by sensors without user intervention. This personal information collection program gathers vast volumes of data concerning the elderly, with an aim to identify habits that are deemed important in the vulnerability model of City4Age to cause potential corrective responses. This deals, in particular, with all data that may be collected via a sensing system from the local area, both at home and in the surrounding areas. Key forms of data are grouped based on the user's movement type (e.g., sleep, walking, rest), indoor or outdoor location (e.g., an enclosed terrace, residence, shopping mall, hospital, church, highway, park), ambient factors (e.g., temperature, humidity, light), and contact between the user and environment (e.g., home heating/cooling systems, television, merchandise purchased in a store, facilities used for public transport). Data are sent to the City4Age website, where the habits of older people and their divergences are analyzed. Most of the information is transmitted via smartphone or smart box with a single-board computer. It is transformed into a common data structure before it is sent to the shared bridge (Fig. 6.4).

Environmental Factors A variety of low-level devices broadcast data composed and presumptuous to the RS by the ESB to monitor environmental factors or identify the occurrence of particular actions. Information is processed to generate events or alerts targeted to a family unit or caregivers. A problem concerns the complexity of the knowledge that the network interfaces supports. A low-cost hoarder, named elegant entry, functions as a translucent concentrator to determine this restriction. This information is gathered, packed, and sent via ESB to the RS, generally via REST crossing points.

Hardware The archetypes used to produce the wearable device are shown in the testing scenario (Fig. 6.5a and b) along with the smart entryway (Fig. 6.5c). The

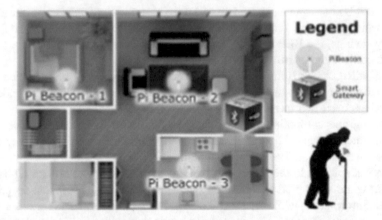

Fig. 6.4 Interior investigation development

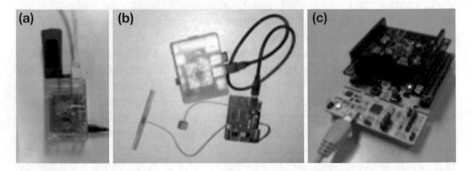

Fig. 6.5 (**a, b**) Smart entryway and (**c**) wearable appliance

elegant entrance, an information hoarder build utilizing a cost low distinct board computer like a Raspberry Pi [4], is the key actor of the scenario sensing middleware.

It carries out a variety of functions, such as compiling information, broadcasting information, and forwarding directives, with different interfaces for both output and input. The standard BT component is worn to communicate with all standard BT apparatus, whereas the BLE component is worn to coordinate the use of all BLE apparatus; in addition, there is an NFC card reader. Two production interfaces are used to transmit the collected information: a wi-fi unit to connect to the Internet and an Ethernet interface mounted on the Raspberry Pi panel.

If the person wants to sit in a chair simply he will send the message through wrist band associated with the centralized IoT system associated with the home. Figure 6.5b shows the initial version designed for use in indoor and outdoor locations, using both Odroid [19] and Arduino [20] devices with BLE, GPS, and GPRS components. This wearable system uses GPS technology for outdoor localization and BLE technology for indoor localization. The next iteration of the model was able to recognize certain incidents, such as falls, using a nucleon panel [15] and

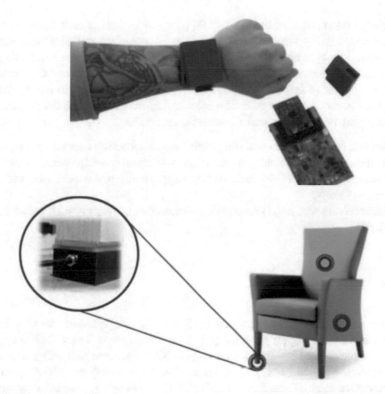

Fig. 6.6 We-care trial products

MEMS and BLE guards [29]. The MEMS guard has several sensors, including a gyroscope, inertial modules, accelerometer, and atmospheric sensors. Figure 6.6 shows the first We-Care archetype, the We-Watch wristband, smart chair, a We-Care floorboard, and the We-Watch gateway. The Web portal and Internet software are not shown.

6.3   We-Watch Wristband

The We-Watch wristband uses a Texas Instruments sensor card in a low-powered expansion of a composed platform of the CC2650 MCU [30] and with numerous MEMS sensors onboard. The multicriterion MCU over IEEE 802.15.4 supports Bluetooth LE 4.0 and 6LoWPAN (2.4 GHz) and is built on Contac-OS, an IoT operating system (OS). The sensor card is a perfect alternative for organizing and inspecting the We-Care wristband due to its condensed size and low-power operation.

They do not have Help IP connectivity and IoT stacking via Contac-OS [31] considering the other available instruments, such as the eZ430-Chronos. The

Contac-OS was used over the 6LoWPAN protocol to provide complete support for the IoT stack. Every We-Watch wristband can accumulate statistics from the sensors, such as body temperature, load, dampness, luminosity, and gesture strength.

Vibrant information unruffled throughout the Withings watch and GPS information and qualitative information physically penetrated addicted to the digital logbook such as mood data; medication information enable the base liner of the Care Receivers "as-is" activities and a start off to recognize:

- Care receivers who are escalating their substantial action / social connections to intentionally communicate, to encourage and reward positive performance.
- Care receivers potentially in danger, to trigger additional supervision and deliverance of alerts.
- Care receivers who need proactive involvement or assistance, such as in response to a fall.

6.3.1   We-Care Panel

A Texas Instruments simple connection CC3200 launch pad strip was used to construct the We-Care digital services and cloud access point. The CC3200 chip-on-chip is composed of an ARM Cortex-M4 [32] CPU core and integrated Wi-Fi connectivity. We used TI-RTOS, an actual time microcontroller unit of operations, for the device stack. Texas Instruments RTOS allows more rapid improvement by removing the requirement to write and sustain system software for developers. The TI-RTOS provides TCP/IP and TLS/SSL piles embedded on behalf of internet interactions, HTTP servers, and Internet protocols.

We-Care can be used as a Wi-Fi dot link, allowing several devices to be connected to and accessed from the same network services, such as a mobile application for the caretaker or in linking the We-Care network to the Internet and the cloud. A concierge can review wristband data recorded in the system using the We-care station profile board and monitor them remotely. The We-Care board runs a web server for remote client connections, with a focus on Port 80. The web application and data files are stored on an SD card attached to the monitor; for example, logs contain the information collected from the wristbands.

6.3.2   We-Watch Access

Because the CC3200 only supports IEEE 802.11 wireless communication, the We-Care panel uses a 6LoWPAN [33] setup via a transceiver attuned with IEEE 802.15.4, much like the CC2538/CC2650. 6LoWPAN access is via a Contac-OS UDP program that has a socket for every wristband in the association, transmitting each of the IPv6 packets inward to the We-Care panel. On the UDP Port 3000, the UDP server monitors and permits connections from isolated users on Port 3001.

6.3.3   Station with Wireless Charging Port

The development of the wi-fi operating network complements the We-Care system, using TIDA-00882 as the frame of reference [4, 34, 35]. This platform facilitates charging of the We-Watch battery because that is what it takes requires support station charging wirelessly while mounted up the base dais. A trouble-free charge system helps aged ready the We-Watch exclusive of having to have cables or complicated systems, link.

6.3.4   Design Applications

Figures 6.7, 6.8 and 6.9 illustrate the stack of We-Care apps. Four basic layers are represented: software, hardware, online sites, and phones. The hardware stratum signifies the hardware boundary support packages (BSP) [36]. In addition to Contac-OS and TI-RTOS libraries, this layer supplies the various boards/hardware used, including the We-Care ring, We-Care entryway, and the We-Care embark. The software layer consists of the protocol stacks and OS components of TI-RTOS [1] and Contac-OS. They are compliant with both IEEE 802.15.4 and IEEE 802.11, and also have an IP-enabled stack to communicate with applications available.

We only run a UDP client-server program for the Contac-OS service layer to allow the exchange of messages among the We-Watch sites. Along with the TI-RTOS, all online services, protocols, and databases are implemented with an IoT C API that communicates within the application layer with a JS IoT API. The application layer on TI-RTOS creates a local web server to show the status of the device and the results, and it communicates all applicable information for the protocol [37] (Figs. 6.7, 6.8 and 6.9).

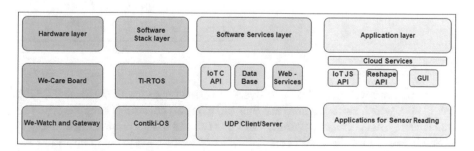

Fig. 6.7 Computer filtering We-Care program

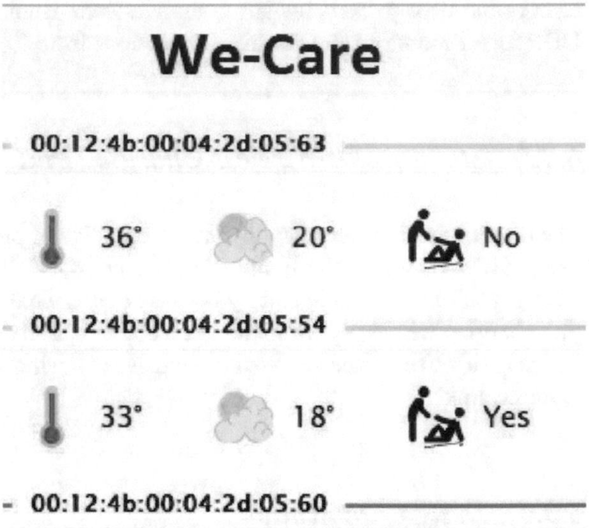

Fig. 6.8 We-Care Internet application: Caregiver app

We-Care

- 00:12:4b:00:04:2d:05:63

36°   20°   No

- 00:12:4b:00:04:2d:05:54

33°   18°   Yes

- 00:12:4b:00:04:2d:05:60

Fig. 6.9 We-Care application: code of warning

We - Care

Distress Button from username was pressed!

Touch to see

1. We-Care web server

The we-care web server is exclusively intended for use with the Bass CC3200 MCU [38], which can accommodate four (up to eight) customers concurrently. The quick deployment offers two primary APIs: IoT.C and IoT.JS printed respectively in C and in JavaScript. Such APIs are interactional to resolve the communication criteria for the machine-to-machine interaction.

2. Web application for We-Care

The We-Care web application is intended to address the system's low-capacity requests. It loads only the necessary data and supports sleep mode to increase energy efficiency, resulting in extended battery life. This program also incorporates an interface to facilitate the interaction of the developer through the We-Care system [39].

It can be written in both HTML5 and CSS3 to be scuttled without the need to add additional applications or plug-ins in any browser and computer. Instead of the SPI flash module on the desktop, request files are accessed from the primary SD card via an API for the file system. This allows for easy modification of the application as the new files are copied to the SD card and access log data is saved to the card where necessary.

Figure 6.5 shows an example of a simple interface that displays the data gleaned from We-Watch bands, such as body and ambient temperatures. In addition, the device event identification is shown, which alerts the caregiver to an incident such as an abrupt fall.

The color outline adjacent to the bracelet identification indicates whether the system is available offline (red) or online (green). The IPv6 uses a standard name for the bracelet, which may be customized as preferred to aid in bracelet recognition.

When the button is pushed for a distress signal, the application shows a caution message accompanied by a audible alarm to inform the caregiver immediately. Additional warnings and notifications may be programmed to guide destinations like the conscientious individual (instead of the caretaker) and urgent situations where the user needs immediate treatment.

3. Protection of data from unauthorized viewers

Data should be protected from unauthorized viewers who may be able to capture IEEE 802.15.4 frames, modify them with false information, or compromise device comportment across the board. The 802.15.4 standard provides cryptographical protection packages for the confidentiality and integrity of algorithms such as the Advanced Encryption Algorithm (AES) [40].

To secure wireless communication between the We-Watch wristband and the We-Wait portal, the We-Watch uses Contac-OS's protection modules AES-128 on-chip and IEEE 802.15.4 (LLSEC) Link-Layer encryption library [41]. The software is built on a security library with layers linked to the establishment of pair-sided keys that implement the Easy Broadcast Protocol for Encryption and Authentication (EBEAP) and the Adaptable Pairway Key Protocols of the establishment scheme (APKES).

We-Watch Battery Lifespan
The We-Watch experimental assessment was carried out to calculate the power used by the We-Watch bracelet for different modes of maneuvering [42]. Battery life can be estimated from the results obtained. Although this MCU supports modes for sleep and deep sleep and Contac-OS is proficient at operating, they are not allowed; therefore, no such information (NA) is available. Table 1 contains all the

measurements and their maximum duration. Because the program requires functionality such as taking samples from sensors, the significance exchange for the We-Watch gateway is 'always on.'

1. Idle: This mode has the lowest real consumption level and is active in most of the time to save electricity. In Idle mode, the contact is off, but can be enabled if the We-Watch has to interact with the message exchange gateway or for maintenance of the network. For example, if sudden motion is recognized by the accelerometer (which might imply a reduction) or the panic button is pressed, the CPU will be interrupted; at this point, We-Watch begins communication and transmits to the We-Care board immediately.
2. Sensors ON: This mode shows the routine sampling of sensors. All sensors take 100 ms to read and absorb a mean current of 6.47 mA.
3. TX Mode: The WeWatch transmits data from the portal to We-Watch after collecting from the sensors. The average duration of this operation is 25.77 mA and not more than 5 ms.
4. RX Mode: This mode occurs after the data are sent. The We-Watch stays 0.3 ms at the gateway in this mode, waiting for a recognition response before returning to Phenomenon [43]. On average, the actual measured consumption is 33.12 mA in this mode. Such communications are required to identify contact failure with the We-Care board.

Power Mode	Description	Time (ms)	Current (mA)
4	RX Mode	33.12	0.3
3	TX Mode	25.77	5
2	Sensors ON	6.47	100
1	Idle	0.76	-
0	Sleep Mode	NA	NA

Current Composition of Power Modes

Contact and 30-s sampling time (where the We-Watch conducts all of the above operations) draws on average 7463 mA for the Sensor ON, TX, and RX modes [44] in 0.105 and 0.760 mA in 29.895 for Idle mode. This results in the current average consumption of 0.784 mA. The approximate battery life is calculated as shown in Fig. 6.10 for a typical 240-mAh rechargeable coin-cell battery.

The expected lifetime of the battery is approximately 306.12 h or 12 days without replacement or re-charges. This gives the user enough time to connect a charger or We-watch bracelet, while retaining services and systems characteristics. When

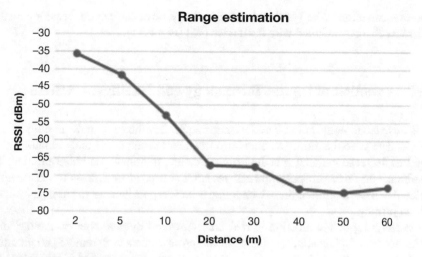

Fig. 6.10 Estimation of range test
batteryli f etime = battery capacity/total consumption
batteryli f etime = 240 mAh/0.784 mA
batteryli f etime = 306.12 h

Fig. 6.11 Summary of the
mobile phone call method

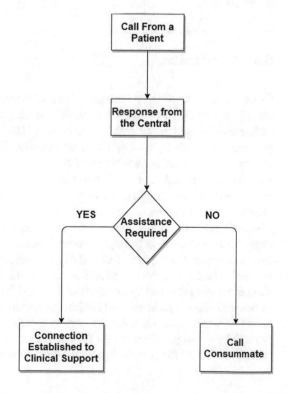

the user can remove the We-Watch independently, he or she can do it easily with the wireless charger included with the device [45] (Fig. 6.11).

6.4 Function of Digital Technology and Intelligent Systems

Assistive technology, home-based monitoring for healthcare, and the intelligent use of big data have the ability to transform treatment in the home and community, as well as decrease national spending on healthcare, treatment, and convalescing. To capitalize on these prospects, further research is needed to address the challenges related to the adoption of these technologies and responsiveness to public questions about privacy.

Performing data collection in real-time, powered by wearable technology and other modes of telehealth, assists healthcare professionals with remote patient monitoring and treatment. In addition, the massive datasets can be used in future studies in numerous fields and are likely to assist progress in prevention and early interventions.

Technologies such as alarms, home networks, tracking systems, and GPS locators may assist caregivers in locating individuals with dementia, but there could be legal concerns. However, caregivers are already using similar products for such purposes.

6.5 Conclusion

Customized, predictive, and integrated IoT approaches can be used to improve modern healthcare delivery. We-Care, an IoT-based healthcare system, can track and gather crucial data about elderly individuals. The work carried out to date uses a proximity and locality data architecture and pilot design to identify data that drives programs to increase the autonomy of older people. Inconspicuous technologies are used to capture and analyze individual data that influences activities, stimulates mental and physical activity, and improves social contact for the elderly population, individuals with mental or physical disability, and individuals with chronic disease. The device design is capable of continuously monitoring activity, movement speed, sleep, and location data through sensors. Data is collected and transmitted in real time via an app that has been safely designed and integrated with smart devices. The predictive healthcare system of IoT H2U can provide early diagnosis and identify dangerous symptoms early on to avoid the need for hospital admission. The length of hospital stays is reduced, and healthcare providers can track patients on a report-based information server using sensors in real time. This also allows the patient to potentially avoid problems while they are alone in their home. IoT device interaction is very cost-effective, with a high degree of security in their communication.

References

1. Islam, S. M. R., Kwak, D., Kabir, M. D. H., Hossain, M., & Kwaki, K. S. (2015, June). The Internet of things for health care: A comprehensive survey. *IEEE Acces, 3*, 678–708.
2. Corchado, J. M., Bajo, J., & Abraham, A. (2008). GerAmi: Improving healthcare delivery in geriatric residences. *IEEE Intelligent Systems, 23*(2), 19–25.
3. Lombardi, A., Ferri, M., Rescio, G., Grassi, M., & Malcovati, P. (2009, October). Wearable wireless accelerometer with embedded FallDetection logic for multi-sensor ambient assisted living applications. *IEEE Sensors, 25–28*, 1967–1970.
4. Yong Lin, Xingjia Lu, Fang Fang, & Jianbo Fan (2013). *Personal health care monitoring and emergency response mechanisms*. In First International Symposium on future information and communication technologies for Ubiquitous HealthCare, Jinhua, 1–3 July 2013, pp. 1–5.
5. Kantoch, E., Augustyniak, P., Markiewicz, M., & Prusak, D. (2014). *Monitoring activities of daily living based on wearable wireless body sensor network*, In36th annual international conference of the IEEE Engineering in Medicine and Biology Society (EMBC), Chicago, IL, 26–30 August 2014, pp. 586–589.
6. Meinel, L., Findeisen, M., Heß, M., Apitzsch, A., & Hirtz, G. (2014). *Automated real-time surveillance for ambient assisted living using an omnidirectional camera*. In IEEE International Conference on Consumer Electronics (ICCE), Las Vegas, NV, 10–13 January 2014, pp. 396–399.
7. Khawandi, S., Daya, B., & Chauvet, P. (2012). *Integrated monitoring system for fall detection in elderly*. In Audio, Language and Image Processing (ICALIP), 2012 International Conference on IEEE, pp. 62–67.
8. Suntiamorntut, W., Charoenpanyasak, S., & Ruksachum, J. (2011). *An elderly assisted living system with wireless sensor networks*. In 2011 4th Joint IFIP Wireless and Mobile Networking Conference (WMNC 2011), October 2011, pp. 1–6.
9. Arboleda, J., Aedo, J., & Rivera, F. (2016). *Wireless system for supporting home health care of chronic disease patients*. In 2016 IEEE Colombian Conference on Communications and Computing (COLCOM), April 2016, pp. 1–5.
10. Kumar, M. (2014). *Security issues and privacy concerns in the implementation of wireless body area network*. In Proceedings of International Conference on Information Technology, Bhubaneswar, Odisha, India, December 2014, pp. 58–62.
11. Lin, C.-C., Chiu, M.-J., Hsiao, C.-C., Lee, R.-G., & Tsai, Y.-S. (2006, October). Wireless health care service system for elderly with Wementia. *IEEE Transactions on Information Technology in Biomedicine, 10*(4), 696–704.
12. Stefanov, D. H., Bien, Z., & Bang, W.-C. (2004). The smart house for older persons and persons with physical disabilities: Structure, technology arrangements, and perspectives. *IEEE Transactions on Neural Systems and Rehabilitation Engineering, 12*(2), 228–250.
13. Banuleasa, S., Munteanu, R., Rusu, A., & Tont, G. (2016). *Iot system for monitoring vital signs of elderly population*. In Electrical and Power Engineering (EPE), 2016 International conference and exposition on IEEE, pp. 059–064.
14. Ausmeier, B., Campbell, T., & Berman, S. (2012). *Indoor navigation using a mobile phone*. In African conference on Software Engineering and Applied Computing (ACSEAC), Gaborone, 24–26 September 2012, pp. 109–115.
15. Aguiar, B., Rocha, T., Silva, J., & Sousa, I. (2014). *Accelerometer-based fall detection for smartphones*. In 2014 IEEE International Symposium on Medical Measurements and Applications (MeMeA), 11–12 June 2014.
16. Mainetti, L., Mighali, V., & Patrono, L. (2015). A software architecture enabling the web of things. *IEEE Internet of Things Journal, 2*(6), 445–454.
17. Ahn, H. S., Datta, C., Kuo, I.-H., Stafford, R., Kerse, N., Peri, K., Broadbent, E., & MacDonald, B. A. (2015). *Entertainment services of a healthcare robot system for older people in private and public spaces*. In Automation, Robotics and Applications (ICARA), 2015 6th International Conference on IEEE, pp. 217–222.

18. Zhongna, Z., Xi, C., Yu-Chia, C., Zhihai, H., Han, T. X., & Keller, J. M. (2009). *Video-based activity monitoring for indoor environments*. In IEEE International symposium on circuits and systems, Taipei, 24–27 May 2009, pp. 1449–1452.
19. Fahim, M., Fatima, I., Lee, S., & Lee, Y-K. (2012) *Daily life activity tracking application for smart homes using android smartphone*. In IEEE 14th International Conference on Advanced Communication Technology (ICACT), pp. 241–245.
20. Villarrubia, G., Bajo, J., de Paz, J. F., & Corchado, J. M. (2014). Monitoring and detection platform to prevent anomalous situations in home care. *Sensors, 14*(6), 9900–9921.
21. Hong, K. S., Bang, O. Y., Kang, D. W., Yu, K. H., Bae, H. J., Lee, J. S., Heo, J. H., Kwon, S. U., Oh, C. W., Lee, B. C., Kim, J. S., & Yoon, B. W. Stroke statistics in Korea: part 1, epidemiology and risk factors: a report from the Korean stroke society and clinical research center for stroke. *Journal of Stoke, 15*.
22. Souza, M. D., Ros, M., & Karunanithi, M. (2012). *An indoor localisation and motion monitoring system to determine behavioural activity in dementia afflicted patients in aged care*. Spec. Issue Aged Care Information 7.
23. Arteaga, S., Chevalier, J., Coile, A., Hill, A. W., Sali, S., Sudhakhrisnan, S., & Kurniawan, S. (2008). *Low-cost accelerometry-based posture monitoring system for stroke survivors*. In: ASSETS'08: Proceedings of the 10th International ACM Sigaccess conference on computers and accessibility, pp. 1–2.
24. Degan, T., Jaeekel, H., Rufer, M., & Wyss, S. (2003). *Speedy: A fall detector in a wrist watch*. In: Proceedings of the Sevent IEEE international symposium of wearable computers.
25. Lustrek, M., Gjoreski, H., Kozina, S., Cvetkovic, B., Mirchevska, V., & Gams, M. (2011). *Detecting falls with location sensors and accelerometers*. In: Proceeding of the Twenty-third innovative applications of artificial intelligence conference, pp. 1662–1667.
26. Chaudhary, A., & Peddoju, S. K. (2018). *The role of IoT-based devices for the better world* (pp. 299–309). Singapore: Springer.
27. Coetzee, L., & Eksteen, J. (2011). *The Internet of things - Promise for the future? An introduction*. In IST-Africa conference proceedings, May 2011, pp. 1–9.
28. Cerf, V., & Senges, M. (2016, February). Taking the Internet to the next physical level. *Computer, 49*(2), 80–86.
29. Raspberry Pi. [Online]. Available: https://www.raspberrypi.org/. Accessed 13 Jan 2016.
30. Odroid. [Online]. Available: http://www.hardkernel.com/. Accessed 13 Jan 2016.
31. Arduino. [Online]. Available: https://www.arduino.cc. Accessed 13 Jan 2016.
32. Nucleo Board (STMicroelectronics). [Online]. Available: http://www.st.com/nucleoF401RE-pr. Accessed 13 Jan 2016.
33. Bluetooth low energy expansion board (STMicroelectronics). [Online]. Available: http://www.st.com/web/catalog/tools/FM116/SC1075/PF260517 Accessed 13 Jan 2016.
34. Attal, F., Mohammed, S., Dedabrishvili, M., Chamroukhi, F., Oukhellou, L., & Amirat, Y. (2015). Physical human activity recognition using wearable sensors. *Sensors, 15*, 31314–31338.
35. Kwapisz, J. R., Weiss, G. M., & Moore, S.A. (2011). *Activity recognition using cell phone accelerometers*. In Proceedings of the 17th Conference on knowledge discovery and data mining, San Diego, CA, USA, 21–24 August 2011, Volume 12, pp. 74–82.
36. Doyle, A., & Mahmood, H. (2016). *Case study' healthy ageing and risk stratification using electronic frailty index Birmingham pilot*.
37. Jovanov, E. (2006) *Wireless technology and system integration in body area networks for m-Health applications*. In Proceedings of 27th annual international conference of the IEEE engineering in medicine and biology, Shanghai, China, January 2006, pp. 7158–7160.
38. Huo, H., Xu, Y., Yan, H., Mubeen, S., & Zhang, H. (2009). *An elderly health care system using wireless sensors networks at home*. In Proceedings of third international conference on sensor technologies and applications, Athens, Greece, June 2009, pp. 158–163.

39. Boutayeb, A., & Boutayeb, S. (2005, January). The burden of non communicable diseases in developing countries. *International Journal Equity Health, 4*, 2–8. https://doi.org/10.1186/1475-9276-4-2.
40. Mattimore, T. J., Wenger, N. S., Desbiens, N. A., Teno, J. M., Hamel, M. B., Liu, H., Califf, R., Connors, A. F., Lynn, J., & Oye, R. K. (1997). Surrogate and physician understanding of patients' preferences for living permanently in a nursing home. *Journal of the American Geriatrics Society, 45*, 818–824.
41. Chakraborty, S., Ghosh, S. K., Jamthe, A., & Agrawal, D. P. (2013). *Detecting mobility for monitoring patient with Parkinson's disease at home using RSSI in a wireless sensor network*. In: The 4th international conference on ambient systems, networks and technologies, the 3rd International conference on sustainable energy information technology, vol. 19, pp. 956–961.
42. Monaci, G., & Pandharipande, A. (2012). *Indoor user zoning and tracking in passive infrared sensing systems*. In Proc. of European Signal Processing Conference (EUSIPCO), Bucharest, 27–31 August 2012, pp. 1089–1093.
43. Reyes-Ortiza, J., Oneto, L., Sam, A., Ghio, A., Parra, X., & Anguita, D. (2016, January 1). Transition-aware human activity recognition using smartphones. *Neurocomputing, 171*, 754–767.
44. Karantonis, D. M., Narayanan, M. R., Mathie, M., Lovell, N. H., & Celler, B. G. (2006). Implementation of a real-time human movement classifier using a triaxial accelerometer for ambulatory monitoring. *IEEE Transactions on Information Technology in Biomedicine, 10*, 156–167.
45. Mainetti, L., Patrono, L., & Rametta, P. (2016). *Capturing behavioural changes of elderly people through unobtrusive sensing technologies*. In 2016 24th international conference on software, telecommunications and computer networks (SoftCOM), Split, Croatia, pp. 1–3. https://doi.org/10.1109/SOFTCOM.2016.7772126.
46. Yacchirema, D. C., Sarabia-Jácome, D., Palau, C. E., & Esteve, M. (2018). *A smart system for sleep monitoring by integrating IoT with big data analytics* (p. 1). IEEE Access.
47. Robie, K. (2010). Falls in older people: Risk factors and strategies for prevention. *JAMA, 304*(17), 1958–1959.
48. Jrad, R. B. N., Ahmed, M. D., & Sundaram, D. (2014). *Insider Action Design Research a multi-methodological Information Systems research approach*. In 2014 IEEE Eighth international conference on Research Challenges in Information Science (RCIS). Marrakech, pp. 1–12.

Chapter 7
Smart Wearable Devices for Remote Patient Monitoring in Healthcare 4.0

U. Hariharan, K. Rajkumar, T. Akilan, and J. Jeyavel

7.1 Introduction

The Internet of Medical Things (IoMT) is a collection of health-related applications with integrated devices, in which healthcare information technology (IT) devices are linked through online computer system networks. Health products built with wi-fi permit machine-to-machine interactions, which is the idea behind the IoMT. IoMT devices connect to the cloud through services such as Amazon Web Services, Microsoft Azure Cloud, Google Cloud Computing, or customized solutions that record data for analysis and storage [1]. IoMT is known as "smart" healthcare; its structural functionality is shown in Fig. 7.1.

Applications of IoMT include the remote monitoring of individuals who have long-term problems and chronic disease. This service monitors treatment and medication adherence, as well as patients' health data through wearable devices that can transmit information to a specific caregiver. IoMT smart devices use sensors and actuators to recognize and evaluate an individual's health data, then transfer the information to the cloud, where it is both accessible to the network and secure. An intelligent system plays a significant part in offering the brain's presence and comfort to doctors and patients. It comprises a method that communicates between programs, apps, and products to allow physicians to monitor patients' vital health

U. Hariharan (✉) · K. Rajkumar
Department of Information Technology, Galgotias College of Engineering and Technology, Noida, Uttar Pradesh, India

T. Akilan
Department of Computer Science and Engineering, Galgotias College of Engineering and Technology, Noida, Uttar Pradesh, India

J. Jeyavel
Department of Electronics and Telecommunication Engineering, Bharati Vidyapeeth College of Engineering, Navi Mumbai, Maharashtra, India

© The Author(s), under exclusive license to Springer
Nature Switzerland AG 2021
D. J. Hemanth et al. (eds.), *Internet of Medical Things*, Internet of Things,
https://doi.org/10.1007/978-3-030-63937-2_7

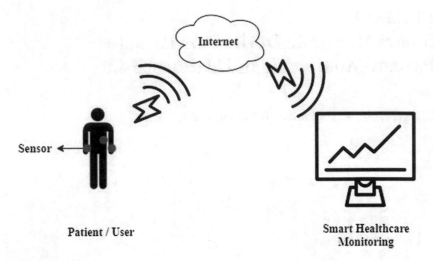

Fig. 7.1 IoMT structural functionality

information. Several products help to monitor useful metrics: wearable health and fitness rings, health and fitness shoes, radiofrequency identification (RFID) pendant watches, along with digital cameras. Additionally, smartphones can provide regular alerts and crisis expertise.

These interconnected Internet of Things (IoT) gadgets provide large amounts of data that must be handled efficiently by vendors. IoT analysis can be carried out to help process such expansive data. A piece of essential information can be made more beneficial with the ability to restore health and well-being when techniques such as data removal and data analytics are used. By 2020, more than 50–55% of techniques used to analyze raw data make better use of this particular arrival of data created by using instrumented machines and applications [2]. Thus, to address privacy concerns, the IoT depends upon several innovative developments. Real-time data from various sources need to be exceptionally easy and quick to access using IoMT. IoMT is managed by smart sensors, which effectively determine, monitor, and assess a selection of overall health signs [3], including heart rate, blood pressure, glucose level, and blood oxygen level. Smart sensors are frequently integrated into drugs and pill bottles, which can provide alerts on whether a person has taken his or her prescribed dose.

Improvements are occurring in IoT's leading solutions. The means of communication with different widgets and the display of data are alerts are changing. Wellness-related solutions are collecting, capturing, sharing, and analyzing user data efficiently in real time. Furthermore, knowledge on the systematic development and advancement of IoT could increase effectiveness. For example, different regenerative units like well-being, social occasions, monitoring the developments, prescription limits, and success are beginning to have a proper sensor held up in them, creating unrefined details, evaluating, and storing it immediately, which enable doctor's to induce appropriate action.

With the assistance of IoT, there are now ways to obtain real-time information from a variety of individuals using smart gadgets that are interconnected within society. Healthcare providers can diagnosis, provide prognosis, and make critical recommendations in a more direct way [5]. This particular evolution provides reliable and useful information, as well as time and cost savings for both patients and providers.

This chapter proposes an IoT device to track an individual's health data, including electrocardiogram (ECG), blood pressure, and heart rate. The collected data can be transferred to caretakers or healthcare providers, imparting a quick and accurate picture of a patient's health.

7.1.1 An Overview of Available IoMT Technology

IoMT devices provide real-time interventions and interconnectivity that can change the healthcare experience, affordability, delivery, and dependability. "Smart" products are connected to other devices or networks can interact with the internet [1]. According to the definition, a device with the following attributes can utilize the IoT:

- Distinctively recognizable.
- Easily discoverable.
- Able to send and receive a response from another device.
- Able to perform simple operations.

The progress in IoT devices is related to the development of sensors, which are more sensitive, easier to discover, and more affordable than in the past. The technology used for the communication of smart devices includes a mix of large wireless networks such as Bluetooth, near filed communication (NFC), wireless sensor network (WSN), low-power wide-area (LPWA), and Long Range [5]. The development of radiofrequency identification (RFID) has also helped to improve IoT applications. This engineering uses microchips for remote data and interaction and is truly a remarkable modification of standard barcodes [3], as it provides both browsing and creation capabilities [4]. Finally, cloud computing is crucial to process all of the (massive) data, which may include temperature, stress, altitude, movement, proximity, sound, or biometrics, among many others.

Wearable smart devices have the potential to greatly improve the understanding, delivery, and results of healthcare and treatment by monitoring an individual patient's health. Nevertheless, challenges exist with the use of sensors for medical trial conduct, information management, and data interpretation using wearables. Wearable devices allow for remote patient monitoring (RPM) of medical data, including heart rate and blood pressure, which can be reviewed remotely by experts and caretakers. The devices can also alert users to potential problems. For example, if an individual is experiencing breathing problems, a notification could be routed to a person monitoring the gadget, who could then appropriately respond to the issue. Furthermore, a wearable device could prevent unnecessary appointments, whereby

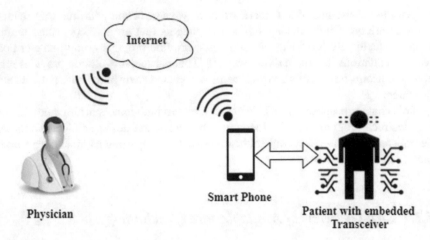

Fig. 7.2 RPM framework

individuals do not need to travel to a physician's office for routine collection of vital signs, which can now be monitored remotely. Hospital staff members can also audit data gathered by the wearable for accuracy.

Remote patient monitoring (RPM, also known as e-health) is a kind of ambulatory therapeutic assistance that allows an individual to use a portable gadget to forgo routine examinations and instead remotely transmit the relevant data to a healthcare provider as needed. RPM includes daily testing devices, such as glucose meters for individuals with and diabetes or heart monitors for aerobic exercise, which are usually routed to a physician's office through an e-health framework. Figure 7.2 shows a unique program that can be used on an individual's internet-connected computer, mobile phone, or even tablet.

7.1.2 Contribution of the Study

Wearable products can monitor a patient and capture real-time data on one's physical state and movements. Wearable sensor-based health and fitness monitors comprise distinct types of adaptable receptors that can be incorporated in textiles, garments, rings, or even placed directly on the body. The receptors can obtain physical data including electrocardiogram (ECG), electromyogram (EMG), pulse rate (PR), body temperature, electrodermal action (EDA), arterial saturation (SpO_2), blood pressure levels (BP), and respiration rate (RR) [5]. Additionally, micro-electro-mechanical program (MEMS)-based monitoring receptors, such as accelerometers, gyroscopes, and magnetic area receptors, are broadly utilized. Constant monitoring of biological indicators may quickly identify certain aerobic, pulmonary, and neurological illnesses at their early onset.

The data received from the receptors inside a wireless body sensor network are transmitted to a neighboring node for processing, preferably via low-power and short-range wireless communications such as Bluetooth, ZigBee, and near field communications (NFC) [6]. The processing node may be a personal digital assistant (PDA), smartphone, computer, custom-made microcontroller, or Field Programmable Gate Array (FPGA). Noninvasive, nonintrusive receptors are essential aspects of long-term smart health monitoring. Wearable receptors are becoming much more comfortable and less obtrusive; thus, they are suitable for checking a person's health condition without disturbing their daily activities. The receptors can gauge several biological signals/parameters, movement, and activity of a person by putting them on numerous places in the body. The development contains low power, compact wearables (sensors, actuators, antennas, sensible textiles), low-cost storage, and computing products accompanied by contemporary correspondence solutions. It reduces the processing cost.

7.2 Wearable Smart Devices for Remote Healthcare Monitoring

Noninvasive, nonintrusive sensors are essential aspects of long-term health and fitness monitoring methods, as described in this section [7].

7.2.1 Body Temperature Monitoring

Body temperature is one of the most important indicators of health issues. Body temperature variations are associated with infection, many inflammatory conditions, malignancy, sleep disorders [8], and menstruation [9]. Coyne et al. [10] noted a correlation between body temperature and initial stroke severity, infarct size, and death rate in stroke patients, with infarct size increasing by 16 mm with a 1 °C increase in body temperature. Additionally, a correlation between cognitive function and body temperature was also noted in the literature [11].

Boano et al. [12] demonstrated an inconspicuous, wireless body temperature monitor that can be used by a person for an extended period of time. Two sensor devices are connected to the epidermis, which may capture and transmit information to a more powerful body-worn primary device, therefore creating a body sensor network. The primary device sends the information to the system. It communicates with healthcare services across the web. The authors attained a precision of 0.04 °C in a temperature range of 16–42 °C. Additionally, it device could obtain circadian rhythms. Afterward, the developers created a wireless monitoring method for long-distance runners, calculating the typical daytime temperature [13].

A safe, wearable temperature monitor also has been created for neonates [14]. A negative temperature coefficient (NTC) resistor is used inside a strap made from smooth bamboo fiber. The developers used a flexible and soft bronze-coated nylon strand for the transfer/transit intermediate interface within the strap rather than rigid cables. The device demonstrated a precision of 0.1 °C compared with a regular thermometer. A reliable, constant link between the sensor and fabric can be critical for this specific application.

The device may be incorporated in a smart coat or even belt, which is a non-invasive, long-term monitoring approach that is appealing for elderly individuals. Mansor et al. [15] explained the setup associated with a wireless temperature monitoring technique utilizing industrial sensors. The climate sensor, which has an incorporated ZigBee wireless node, records and sends information to a microcontroller. The microcontroller sends data to a server that is on the remote side along with a WLAN. An ethernet shield for an Arduino microcontroller was used to improve a module. Related heart rate and temperature monitoring methods have been described elsewhere [16].

Sim et al. [17] developed a method of implanting the flux probe for a dual-heat probe, with two dual-sensor thermometers interfaced in a neck pillow. Because the jugular vein passes through the neck, your head has a stable association with the core body temperature (CBT). The temperature calculated by this product is similar to that assessed by an infrared thermometer through the tympanic membrane. The developers recommended a curve-fitting technique to enhance the natural gradual effect period of a dual-heat-flux thermometer. A method also may be created for individuals to use while sleeping, inside the neckline of a shirt, or worn as a fabric strip. Kitamura et al. [18] created a heat sensor pin that gauges CBT over the epidermis. The round steel wire includes two heat flow routes for two distinct resistances; every path has a set of climate transmitters, and the receiver is attached in both ends. This method shows much effective and greater precision (96% association with a measurement calculated out of a null heat flow thermometer) as it tends to be stable or balanced heat. The required time period can undoubtedly be improved with artificial intelligence instruments rather than a copper instrument.

7.2.2   Activity Monitoring

Monitoring a person's aerobic activity and movement may be useful for rehabilitation, sports, early detection of cognitive or musculoskeletal illnesses, or sense of balance evaluation. A person's walking patterns have been associated with their health problems [19]. Any gait abnormalities may be suggestive of potential musculoskeletal or nervous system. For example, individuals in the early phase of neurodegenerative problems, such as Alzheimer and Parkinson disease, often display certain gait abnormalities [20], such as small and shuffled steps with Parkinson disease. In addition, older individuals are more susceptible to falls, which may cause accidents, hip fractures, traumatic brain injuries, and other bone fractures

[20]. Quantitative assessment and analysis can predict numerous illnesses, fall potential, and rehabilitation time after injury.

At home, camera-based methods are helpful for monitoring exercise [21]. However, the fixed location limits the detection of motion to a particular area. Other methods are more expensive and complicated. Recently, wearable movement receptors, such as magnetometers, gyroscopes, and accelerometers, have become more popular for monitoring activity in real time [22]. A sensor evaluates the body's angular and linear activity by selecting certain parameters to measure. A schematic of the activity monitoring with accelerometers and gyroscopes is provided in Fig. 7.3.

Bertolotti et al. [23] created a small, wireless wearable unit used to evaluate balance by monitoring limb motions for an extended period using a gyroscope, a magnetometer, and an accelerometer. Several models may also be attached within a body sensor network to calculate various parameters. The method was authenticated by evaluating the center-of-mass (CoM) movement calculated through a Nintendo Wii Balance Board (https://www.analog.com/en/applications/markets/healthcare-pavilion-home.html). Additionally, analysis is recommended to determine treatment options based on data measurements. Panahandeh et al. [24] recommended a gait evaluation algorithm based upon the Hidden Markov Model with measurements obtained from a tri-axial accelerometer and gyroscope. They proposed a function removal technique grounded on measuring Discrete Fourier Transform (DFT) coefficients are commencing sections of motion. Experts attained higher categorical precision depending on sport function. Because of the lack of wireless connectivity, a computer had to be used for testing or research, which affected the subject's activities. Otherwise, the evaluation was carried out with an ad-hoc schedule. The device can be upgraded by integrating wireless connectivity and including other algorithms.

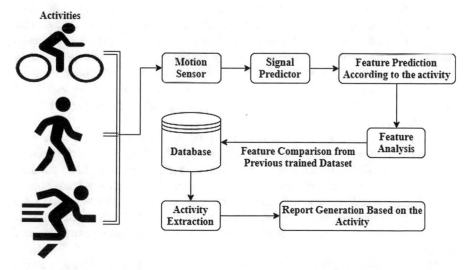

Fig. 7.3 Activity monitoring

Friedman et al. [25] created a wearable wrist and a finger joint checking technique using magnetic technologies. The device is made up of a neodymium band with data storage. The finger band has a magnetic area assessed by two tri-axis magnetometers installed in the wrist device. A radial timeframe feature estimates the characteristics of the wrist and finger bones in the dimensions. The authors found the same result for the angular distance for wrist bones.

Nevertheless, finger flexion and extension evaluation is complicated. Evaluation can be improved by ensuring the calibration is correct for each user. A method for joining perspective evaluation of pairing kinematic arm designs together with the express design algorithm continues to be accessible [26]. Inertial dimensions were taken from three sets of inertial measurement unit (IMUs) placed on the top arm, forearm, and a wrist. The expressive design integrated arbitrary drift clothes areas and zero velocity adjustments, which could reduce the sensor flow. The unit also found restrictions of joint motions with substantial reliability. The authors approximated joint movement using the unscented Kalman filter [27]; a more complex algorithm, the extended Kalman filter (EKF), can be used for devices with increased nonlinearities. It may attain a top-level evaluation and precision during various intensities of arm actions.

Validation of this particular algorithm for the primary subject matter is essential. Kalman-based solutions are computationally rigorous and require a high sampling pace for human activities. Thus, using the devices for real-time programs might be difficult.

7.2.3   Galvanic Skin Response Monitoring

Autonomic nervous system (ANS) responses can be measured with an external or internal stimulus by controlling tasks within its two subdivisions: the sympathetic and parasympathetic nervous systems [28]. The parasympathetic system conserves and restores all of the body's electricity, the sympathetic system sparks the so-called flight-or-fight response by increasing the metabolic changes to cope with an outside stimulus. The sympathetic system can increase pulse rate, blood pressure, and sweat secretion to prepare the body for activity by pumping additional blood to muscles, the lungs, and the brain.

Much is unknown about dysautonomia, postural orthostatic tachycardia syndrome (POTS), and tachycardia. Increased sweat secretion from eccrine glands loads the sweat ducts. Sweat, being a weak electrolyte, increases the skin's conduction through an enhanced discharge of secretion. Perturbation of epidermis conductance is known as EDA. Galvanic skin response (GSR) mirrors the response of the sympathetic central nervous system. It may also offer a straightforward, reliable, and sensitive approach for evaluating sympathetic responses linked to anxiety and emotion [29].

Generally, GSR is calculated from the skin's components, from the many sweat glands in the soles, fingers, or palm. Passive measurement, a DC voltage, has been used on two body electrodes. In addition, the epidermis conductance can be computed by Ohm's law. Early research in this area mostly centered on time-limited GSR measurement devices in the laboratory or healthcare facilities. The advancement of wearable technologies is providing exciting opportunities for inconspicuous, extended-use GSR [30]. Long-term GSR monitoring allows for evaluation of the sympathetic central nervous system for an extended period to possibly obtain important biological data that cannot be captured during short-term monitoring. In addition, wearable GSR allows users to monitor GSR from home to better evaluate their psychophysiological condition versus analysis occurring in hospitals or laboratories with fewer measurements [31]. A diagram of wearable GSR monitoring is provided in Fig. 7.4.

A wearable GSR sensor can carry out measurements from the entire body and transmit information via a Bluetooth device. This sensor is located on a flexible printed circuit board (PCB) that includes silicon, making continuous contact with a curved body area [32]. A conductive polymer foam was also used because the initial substance was only for the flexible electrodes. The electrodes' adaptive dynamics provide a steady and dependable skin-electrode user interface and comfort for the end user. The GSR assessed through this method had a significant association (average ~ 0.768) with the typical GSR process, although the comparison was carried out with a limited number of topics. Garbarino et al. [33] created a multi-sensor wristband (Empatica E3) that incorporated GSR, temperature, Photoplethysmogram (PPG), and a movement sensor. The information acquisition unit has a dimension of 4x4 implanted within a wristband. It can capture information from all four sensors continuously for 30+ hours. The device may also stream data in real-time using Bluetooth. A longer battery life, lower-power wireless connectivity, and ability to track multiple parameters are important characteristics for future development.

Setz et al. [34] attempted to differentiate cognitive stress and load in GSR measurements. The developers individually assessed GSR from the toes for a selection of topics during two synthetically produced mental circumstances mimicking cognitive stress and load. A wearable device calculated and transmitted the GSR signal via Bluetooth to a website. For 16 options available for GSR signal distributions, the correct GSR level was rapidly selected, offering much better outcomes for differen-

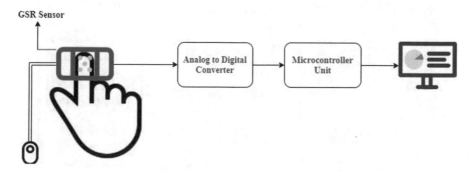

Fig. 7.4 Galvanic skin response (GSR) monitoring

tiating the cognitive load of anxiety. Akbulut et al. [35] examined the feasibility of wearable, wireless GSR and HR monitoring for evaluating the strain amount and the performance of an autonomic process in an ambulatory state.

7.2.4 Multiple Device Monitoring

Nearly all of the previously discussed track one specific biosignal or parameter, such as only heart rate and ECG [36]. However, it is often essential to track multiple physical symptoms, such as pulse or heart rate, respiratory rate, blood pressure, and body temperature. Frequently, both oxygen saturation in the bloodstream and GSR provide a much better evaluation of a person's health issues. However, using a different device for each parameter is neither pragmatic nor ergonomically possible for ambulatory monitoring. A system of several sensors lodged together inside a wearable device with transmitter and receiver components may be a practical option to keeping track of multiple parameters. For example, many researchers have used the Bramwell-Hill and Moens-Korteweg calculations to determine blood pressure from pulse transit time. Figure 7.5 presents the concept of multiple sensor monitoring.

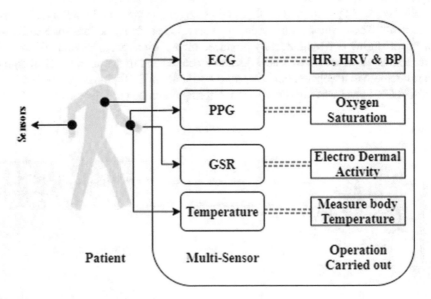

Fig. 7.5 Multiple sensor monitoring

7.3   Fabric-Based Wearables

Smart fabrics can be linked to sensors embedded within, providing a comfortable and unobtrusive way to monitor biological indicators [37]. The active realizing substance is generally constructed on a substrate and is directly connected with the body surface or remains encapsulated within the fabric [38]. Smart textiles can be used with body electrodes to measure electrophysiological indicators such as ECG, electroencephalography, GSR, and EMG. Textile-based electrodes of silver chloride have been reported to be reliable [39]. Textile electrodes can be categorized into two different types: passive and active. Passive textile electrodes determine the electrical qualities of the epidermis. They may be utilized to keep track of cardiac or muscle measurement. They also have been used in GSR measurements, with perspiration being evaluated by connecting electrodes on the skin. Conventional electrodes are affixed to the skin by conductive and adhesive gel , which may cause irritation or allergies [40]. Also, signal quality may decrease when the gel dries. Damp electrodes offer better signal quality, but they are not ideal for long-term use or wearable devices [41].

Fabric electrodes may be sewn to appropriate places on the clothing or may exclusively deposit conductive levels within the cloth. The conductive levels can be created on the clothing surface by depositing nanofibers [42] utilizing an electrode-position procedure or using a conductive level together through assistance to display spluttering procedure by removing and evaporation [43].

7.3.1   Fabric-Based Rectal Sensor

A rectal thermometer is often the most accurate method for determining body temperature. A tympanic thermometer have measurement reliability issues [44], as do oral and axillary (armpit) temperature measurements [45]. Skin temperature varies from the core body temperature by as much as 3 °C [46]. A fabric-based temperature sensor works well for calculating body temperature inside a wearable wedge. It is composed of fiber or yarn using traditional garment production solutions, such as weaving, embroidery, knitting, and then printing. Climate receptors fabricated on flexible substrates also may be integrated into textiles. Sensitivity, linearity, accuracy, and rapid results in a range of 34–40 °C are vital characteristics of these heat sensors. The number of suitable encapsulation substances is also critical to protect it from external physical and environmental impacts.

7.3.2 Fabric-Based Activity Monitoring

Researchers have investigated MEMS-based inertial receptors, including acceler-ometers, magnetic field sensors, gyroscopes, and their combinations to monitor human locomotion. A very small PCB panel may be inserted into belts, flexible rings, and Velcro straps. MEMS-based movement sensors are small and inexpensive, with great detection ability, reliability, and low power requirements; thus, they are ideal for real-time and long-term monitoring. However, rigid PCB boards might not feel comfortable for some people. Rajdi et al. [47] fabricated a MEMS acceler-ometer on cotton fabric to compute pelvic tilt. The accelerometer used the piezoresistive characteristics of conductive Ag nanoparticles incorporated in the garments by pressing/ironing. If the tilt changes, a cantilever beam system on the accelerom-eter suffers from physical tension, ultimately adjusting the conductive content linearly. Researchers found an excellent correlation between distant relative opposi-tion shifts and the stress used along with the textile-based accelerometer.

Stretchable and flexible stress sensors were also used by many researchers in textile-based monitoring. Stress sensors determine physical deformation by modify-ing electric qualities, including capacitance and resistance, in reaction to physical tension. The stress receptors for textile applicationssmust be extremely versatile, stretchable, and long-lasting. Furthermore, sensitivity and a quick response/recov-ery period are vital for real-time monitoring [48].

An optical fiber Bragg Grating (FBG)-based monitoring technique has been described elsewhere [49]. One optical FBG sensor wrapped with polymer foil was built into a flexible joint strip. The flexion motion in knees leads to stress and varies from the resonance wavelength on the FBG. Additionally, it incorporates FBG sensors within a glove and Velcro straps to enhance FBG sensors' ability to detecting finger motions recognizing finger movements and breath out. Rather than optical strength, the FBG sensor carried out measurements based on the wavelength, which was much less sensitive to outside noise and variations within the optical energy source. The system exhibited excessive awareness, measurement accuracy, and stability.

Kaysir et al. [50] created an optical fiber-based versatile pressure sensor that may be incorporated into textiles. The optical fiber was created with an adaptable copo-lymer that contains polyurethane and silicon. The optical fiber has an oval-shaped/egg-shaped twist down the cross-section of its outside pressure. This particular deformation increases light deflection inside the fiber, resulting in decreased sever-ity of gentle within the result. This particular deformation increases light deflection inside the fiber, resulting in decreased severity of the outcome. The power, conse-quently, could be approximated as a result of gentle severeness. The sensor is versa-tile, allowing it to be incorporated into garments to identify, for example, limb motion and respiratory rate. The sensor used here is hypersensitive for temperature, bends, then finally strains, which causes accuracy issues.

7.4   Communication Technologies for Remote Healthcare Monitoring

The biological indicators assessed through the on-body sensors require the ability to transfer information to a remote server. For basic monitoring, a short distance is needed to transfer the calculated information to the closest gateway node, such as a PDA, custom-designed FPGA, laptop, personal computer, or smartphone. A more complex entry must be transmitted a remote server at a clinic. Thus, the information may be transferred via an online network or mobile correspondence system. At present, most cellular networks provide seamless access to the web by way of Long-Term Evolution (LTE) solutions [51]. Nevertheless, strong authentication and encryption are necessary to protect sensitive, private data during transmission.

In short-range correspondence, the sensors can communicate directly with the gateway. A wireless body area network (WBAN) may be created using sensors and topology through primary WBAN nodes. The WBAN node transmits information to the gateway, immediately executing a few processes. The on-body sensors and WBAN node may commincate in a wireless or wired manner. User mobility may be delayed with wired contacts, and also it could cause regular unsuccessful contacts. As a result, wired communication is not ideal for long-term and wearable monitoring methods.

Bluetooth is the preferred low-power radiofrequency technological innovation, as it is prevalent in products such as laptop computers, smartphones, and health and fitness trackers for small amounts of data. It uses a 2.4 GHz frequency band within the ISM stereo spectrum and sends signals to more than 78 specified route, using the frequency-hopping spread spectrum (FHSS) technique for communication. ZigBee is another wireless standard format that requires low power and inexpensive interactions inside small areas. It works on the unlicensed 2.4 GHz frequency rings of the ISM spectrum and sends information to more than 16 channels, using quadrature phase shift keying (QPSK) modulation for interaction.

Medical implant services use low-range, very-low-energy wireless technology to communicate with fixed health-related products such as cardiac pacemakers, defibrillators, and neurostimulators. It works in the frequency band of 402–405 MHz with 300-kHz stations. This particular frequency band provides great signal propagation qualities within the human body, making it ideal for implantable units. There are some different wireless solutions for short-range correspondence, such as IrDA, UWB, RFID, and NFC [6, 52]. Because of its high energy utilization and complicated configurations, wi-fi not suitable to track the patient for a long time period due to the additional power required to extend the network.

7.5 Proposed Methodology

The proposed RPM framework has the ability to filter the individual patient details through the IoT, which could assemble the data with this framework. The system could compile an individual's heart rate, ECG, body temperature, and BP, then transmit an alert to a person's caretaker with his or her existing condition and complete remedial information. The caretaker can examine the patient and provide the appropriate response according to the patient's health status without an in-person healthcare visit. The process uses body receptors that capture details from each sensor and send it to a data server, where the information may be viewed and downloaded by healthcare professionals. Sustaining a data server is crucial to ensure a history of health data is available for the individual. A general and enhanced view of previous health issues is shown in Fig. 7.6.

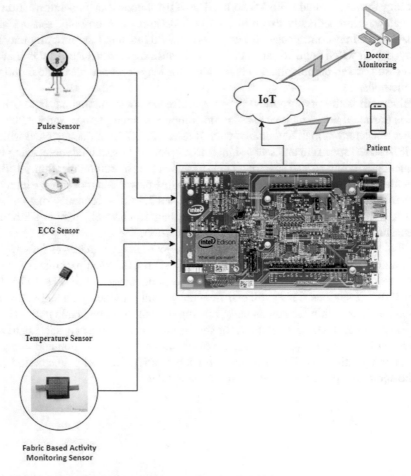

Fig. 7.6 Remote patient monitoring framework using Intel Edison

The framework consists of an Intel Edison (system on chip). This framework receives the information from all of the receptors attached to the individual and uploads this information to the internet server via an ethernet connection. The physician can monitor a patient's details by an internet-based program or mobile device. The sensors connected to inpatient include a heart monitor (SEN 11574 pulse sensor), which requires +3.3 V to +5 V at VCC and a climate sensor that provides a history of the person's general health and fitness. For overall body temperature, an LM 35 climate sensor (DHT 11) is used. If the heartbeat detector is operating, the corresponding LED flashes in unison with every heartbeat. This particular electronic data is usually connected to a microcontroller to determine the beats per minute (BPM).

The system part consolidates an Arduino IDE that is a necessary framework for our Intel Edison Board that was used-to swap the last code of ours of staying in touch with a repository. All information from the receptors is delivered to a Windows, Apache, MySQL, PHP (WAMP)-based data source server to log regular details, which should assist caregivers with determining the need for a specialist consultation and prescriptions. Additionally, these datasets stored in the database are used to plot graphs to identify every sensor's patterns. The server could move the individual's data source with unobtrusive components and patient history. The information on the server can be viewed at any time by an administrator using the internet, An administrator or mobile application can view the server's information, and the service provider can notice the present living feed of the person's healing situation. Tracking the patient's overall health report is reserved for a potential guide for professionals. It could hold and safeguard the patient's details for 24×7 hours of various people. A individual could also view his or her personal health record.

With this display, the professional or a guardian can enter the web page's login credentials, as shown in Fig. 7.7. After the qualifications are confirmed, the device shows a summary of individuals assigned to the healthcare provider. The professional chooses which individual's information to display, including temperature, heart rate, ECG, and more.

Fig. 7.7 Remote patient monitoring login page

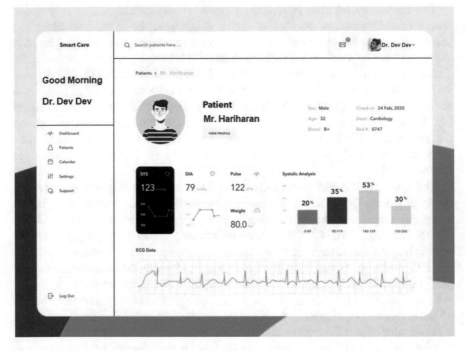

Fig. 7.8 Patient monitoring

To protect patient data, transferred information is encrypted while being delivered to the data source server, then is decoded while duplicated data is copied to the website, as shown in Fig. 7.8.

In picture subheadings, providers glean insight from the patient's current readings when communications with devices complete without error. If the gadget isn't associated or any of the sensors isn't attached to the patient. At that time, each reading should be set to zero to prevent an incorrect analysis. Because of the possibility that the gadget may turn off inadvertently, this web page displays any shut-off errors. The information coming from different receptors is now published to the data server with plot charts and evaluations.

7.6   Conclusion

Continual monitoring of a patient's overall health condition can provide instant insight into an individual's well-being. Wearable sensors and actuators, integrated with sophisticated communications and data solutions, unlock new opportunities for remote healthcare products. The methods consist of monitoring and evaluation with predictive algorithms, which can determine the prognosis of particular illnesses with greater confidence, therefore reducing unnecessary treatment and incorrect diagnoses. A device can generate an alert to notify the user or a healthcare provider of any issues. Smart fabrics, including textile-based connections for sensors, can be used with wearable devices to provide a more comfortable, less intrusive framework for overall health monitoring.

References

1. IoT Healthcare Solutions-Medical Internet of Things for Healthcare. https://medium.com/@ itcubeservice/iot-healthcare-solutions-medical-internet-of-things-for-healthcare-1f9aac81aad6
2. The Digital Universe in 2020: Big Data, Bigger Digital Shadows, and Biggest Growth in the Far East. https://www.emc.com/leadership/digitaluniverse/2012iview/big-data-2020.htm
3. Shahid, N., & Aneja, S. (2017). *Internet of things: Vision, application areas and research challenges.* In 2017 International Conference on I-SMAC (IoT in Social, Mobile, Analytics, and Cloud) (I-SMAC), Palladam, pp. 583–587.
4. He, D., & Zeadally, S. (2015). An analysis of RFID authentication schemes for the internet of things in healthcare environment using elliptic curve cryptography. *IEEE Internet of Things Journal, 2,* 72–83.
5. Stankovic, J. A. (2014, February). Research directions for the Internet of things. *IEEE Internet of Things Journal, 1*(1), 3–9.
6. Ullah, S., Higgins, H., Braem, B., et al. (2012). A comprehensive survey of wireless body area networks. *Journal of Medical Systems, 36,* 1065–1094.
7. Majumder, S., Aghayi, E., Noferesti, M., Memarzadeh-Tehran, H., Mondal, T., Pang, Z., & Deen, M. (2017). Smart homes for elderly healthcare Recent advances and research challenges. *Sensors, 17,* 2496.
8. Agoulmine, N., Deen, M., Lee, J., & Meyyappan, M. (2011). U-Health smart home. *IEEE Nanotechnology Magazine, 5,* 6–11.
9. Lack, L., Gradisar, M., Van Someren, E., Wright, H., & Lushington, K. (2008). The relationship between insomnia and body temperatures. *Sleep Medicine Reviews, 12,* 307–317.
10. Kräuchi, K., Konieczka, K., Roescheisen-Weich, C., Gompper, B., Hauenstein, D., Schoetzau, A., Fraenkl, S., & Flammer, J. (2013). Diurnal and menstrual cycles in body temperature are regulated differently: A 28-day ambulatory study in healthy women with thermal discomfort of cold extremities and controls. *Chronobiology International, 31,* 102–113.
11. Coyne, M., Keswick, C., Doherty, T., Kolka, M., & Stephenson, L. (2000). Circadian rhythm changes in core temperature over the menstrual cycle: A method for non-invasive monitoring. *American Journal of Physiology-Regulatory, Integrative, and Comparative Physiology, 279,* R1316–R1320.
12. Shibasaki, K., Suzuki, M., Mizuno, A., & Tominaga, M. (2007). Effects of body temperature on neural activity in the Hippocampus: Regulation of resting membrane potentials by transient receptor potential Vanilloid 4. *Journal of Neuroscience, 27,* 1566–1575.
13. Boano, C., Lasagni, M., Romer, K., & Lange, T. (2011). *Accurate temperature measurements for medical research using body sensor networks.* In 2011 14th IEEE International symposium on object/component/service-oriented real-time distributed computing workshops.
14. Boano, C., Lasagni, M., & Romer, K. (2013). *Non-invasive measurement of core body temperature in Marathon runners.* In 2013 IEEE International conference on body sensor networks.
15. Chen, W., Dols, S., Oetomo, S., & Feijs, L. (2010). *A monitoring body temperature of newborn infants at neonatal intensive care units using wearable sensors.* In Proceedings of the Fifth international conference on body area networks – BodyNets '10.
16. Mansor, H., Shukor, M., Meskam, S., Rusli, N., & Zamery, N. (2013). *Body temperature measurement for the remote health monitoring system.* In 2013 IEEE International Conference on Smart Instrumentation, Measurement and Applications (ICSIMA).

17. Asaduzzaman Miah, M., Mir Hussain Kabir, Siddiqur Rahman Tanveer, M., Akhand, M. (2015). *Continuous heart rate and body temperature monitoring system using Arduino U.N.O. and Android device*. In 2015 2nd International Conference on Electrical Information and Communication Technologies (EICT).
18. Sim, S. Y., Lee, W. K., Baek, H.J., & Park, K. S. (2012). *A non-intrusive temperature measuring system for estimating deep body temperature in bed*. In Proceedings of the 2012 Annual International Conference of the IEEE Engineering in Medicine and Biology Society, San Diego, CA, USA, pp. 3460–3463.
19. Kitamura, K.-I., Zhu, X., Chen, W., & Nemoto, T. (2010). Development of a new method for the non-invasive measurement of deep body temperature without a heater. *Medical Engineering & Physics, 32*, 1–6.
20. Mulroy, S., Gronley, J., Weiss, W., Newsam, C., & Perry, J. (2003). Use of cluster analysis for gait pattern classification of patients in the early and late recovery phases following stroke. *Gait & Posture, 18*, 114–125.
21. Coutinho, E. S. F., Bloch, K. V., & Coeli, C. M. (2012). One-year mortality among elderly people after hospitalization due to fall-related fractures: Comparison with a control group of matched elderly. *Cadernos de Saúde Pública, 28*, 801–805.
22. Ni, B., Wang, G., & Moulin, P. (2013). RGBD-HuDaAct: A color-depth video database for human daily activity recognition. In *Consumer depth cameras for computer vision* (pp. 193–208). London: Springer.
23. Deen, M. J. (2015). Information and communications technologies for elderly ubiquitous healthcare in a smart home. *Personal and Ubiquitous Computing, 19*, 573–599.
24. Bertolotti, G. M., Cristiani, A. M., Colagiorgio, P., Romano, F., Bassani, E., Caramia, N., & Ramat, S. A. (2016). Wearable and modular inertial unit for measuring limb movements and balance control abilities. *IEEE Sensors Journal, 16*, 790–797.
25. Panahandeh, G., Mohammadiha, N., Leijon, A., & Handel, P. (2013). Continuous hidden Markov model for pedestrian activity classification and gait analysis. *IEEE Transactions on Instrumentation and Measurement, 62*, 1073–1083.
26. Friedman, N., Rowe, J. B., Reinkensmeyer, D. J., & Bachman, M. (2014). The manometer: A wearable device for monitoring daily use of the wrist and fingers. *IEEE Journal of Biomedical and Health Informatics, 18*, 1804–1812.
27. El-Gohary, M., & Mcnames, J. (2015). Human joint angle estimation with inertial sensors and validation with a RobotArm. *IEEE Transactions on Biomedical Engineering, 62*, 1759–1767.
28. Chernikova, O. S. (2018). *An adaptive unscented Kalman filter approach for state estimation of nonlinear continuous-discrete system*. In 2018 XIV International scientific-technical conference on Actual Problems of Electronic Instrument Engineering (APEIE), Novosibirsk, pp. 37–40.
29. Wang, Z., Wang, F., & Ji, X. (2019). *Analysis of autonomic nervous system based on HRV*. In 2019 4th International Conference on Mechanical, Control and Computer Engineering (ICMCCE), Hohhot, China, pp. 309–3095.
30. Bakker, J., Pechenizkiy, M., & Sidorova, N. (2011). *What's your current stress level? Detection of stress patterns from GSR sensor data*. In 2011 IEEE 11th international conference on data mining workshops, Vancouver, BC, Canada, pp. 573–580.
31. Eng, S., Al-Mai, O., & Ahmadi, M. (2018). A 6 DoF, wearable, compliant shoe sensor for total ground reaction measurement. *IEEE Transactions on Instrumentation and Measurement, 67*(11), 2714–2722.
32. Crivello, A., Barsocchi, P., Girolami, M., & Palumbo, F. (2019). The meaning of sleep quality: A survey of available technologies. *IEEE Access, 7*, 167374–167390.
33. Hernando-Gallego, F., Luengo, D., & Artés-Rodríguez, A. (2018). Feature extraction of galvanic skin responses by nonnegative sparse deconvolution. *IEEE Journal of Biomedical and Health Informatics, 22*(5), 1385–1394.
34. Haghi, M., Stoll, R., & Thurow, K. (2019). Pervasive and personalized ambient parameters monitoring: A wearable, modular, and configurable watch. *IEEE Access, 7*, 20126–20143.

35. Setz, C., Arnrich, B., Schumm, J., Marca, R. L., Troster, G., & Ehlert, U. (2010). Discriminating stress from cognitive load using a wearable EDA device. *IEEE Transactions Information Technology in Biomedicine, 14*, 410–417.
36. Akbulut, F. P., Özgür, Ö., & Cınar, İ. (2019). *e-Vital: A wrist-worn wearable sensor device for measuring vital parameters.* In 2019 Medical technologies congress (TIPTEKNO), Izmir, Turkey, pp. 1–4.
37. Tasneem Usha, R., Sazid Sejuti, F., & Islam, S. (2019). *Smart monitoring service through self sufficient healthcare gadget for elderly.* In 2019 IEEE International Symposium on Smart Electronic Systems (iSES) (Formerly iNiS), Rourkela, India, pp. 276–279.
38. Somov, A., Alonso, E. T., Craciun, M. F., Neves, A. I. S., & Baldycheva, A. (2017). *Smart textile: Exploration of wireless sensing capabilities.* In 2017 IEEE Sensors, Glasgow, pp. 1–3.
39. Yu, H., Zheng, Z., Ma, J., Zheng, Y., Yang, M., & Jiang, X. (2017). *Temperature and strain sensor based on a few-mode photonic crystal fiber.* In 2017 IEEE Sensors, Glasgow, pp. 1–3.
40. Takeshita, T., Yoshida, M., Takei, Y., Ouchi, A., & Kobayashi, T. (2019). *Cubic flocked electrode embedding amplifier circuit for smart ECG textile application.* In 2019 20th International conference on solid-state sensors, Actuators and microsystems & eurosensors XXXIII (TRANSDUCERS & EUROSENSORS XXXIII), Berlin, Germany, pp. 2189–2192.
41. Chen, H., et al. (2020). Design of an integrated wearable multi-sensor platform based on flexible materials for neonatal monitoring. *IEEE Access, 8*, 23732–23747.
42. Nemati, E., Deen, M., & Mondal, T. (2012). A wireless wearable ECG sensor for long-term applications. *IEEE Communications Magazine, 50*, 36–43.
43. Kõiv, H., Pesti, K., Min, M., Land, R., & Must, I. (2020). Comparison of the carbon nanofiber-/fiber- and silicone-based electrodes for bioimpedance measurements. *IEEE Transactions on Instrumentation and Measurement, 69*(4), 1455–1463.
44. Cho, G., Jeong, K., Paik, M. J., Kwun, Y., & Sung, M. (2011). Performance evaluation of textile-based electrodes and motion sensors for smart clothing. *IEEE Sensors Journal, 11*(12), 3183–3193.
45. López, L., Domínguez, G. E., Cardiel, E., & Hernández, P. R. (2019). *Wireless measurement system for mean body temperature estimation.* In 2019 Global Medical Engineering Physics Exchanges/Pan American Health Care Exchanges (GMEPE/PAHCE), Buenos Aires, Argentina, pp. 1–6.
46. Caya, M. V. C. et al. (2017). *Basal body temperature measurement using e-textile.* In 2017 IEEE 9th international conference on Humanoid, Nanotechnology, Information Technology, Communication and Control, Environment and Management (HNICEM), Manila, pp. 1–4.
47. Narczyk, P., Siwiec, K., & Pleskacz, W. A. (2016). *Precision human body temperature measurement based on thermistor sensor.* In 2016 IEEE 19th international symposium on Design and Diagnostics of Electronic Circuits & Systems (DDECS), Kosice, pp. 1–5.
48. Rajdi, N. N. Z. M., Bakira, A. A., Saleh, S. M., & Wicaksono, D. H. (2012). Textile-based micro electro mechanical system (MEMS) accelerometer for pelvic tilt measurement. *Procedia Engineering, 41*, 532–537.
49. Baumbauer, C., Ting, J., Thielens, A., Rabaey, J., & Arias, A. C. (2019). *Towards wireless flexible printed wearable sensors.* In 2019 IEEE 8th International Workshop on Advances in Sensors and Interfaces (IWASI), Otranto, Italy.
50. Silva, A. F. D., Pedro, R., Paulo, J., & Higino, J. (2013). Photonic sensors based on flexible materials with FBGs for use on biomedical applications. In *Current trends in short- and long-period fiber gratings* (pp. 105–132). Rijeka: InTech.
51. Kaysir, M. R., Stefani, A., Lwin, R., & Fleming, S. (2017). *Flexible optical fiber sensor based on polyurethane.* In 2017 Conference on Lasers and Electro-Optics Pacific Rim (CLEO-PR), Singapore, pp. 1–2.
52. Bojović, B., Giupponi, L., Ali, Z., & Miozzo, M. (2019). Evaluating unlicensed LTE Technologies: LAA vs LTE-U. *IEEE Access, 7*, 89714–89751.

Chapter 8
Remote Healthcare Monitoring for Parkinson's Disease in a Smart Way

R. Sujatha, T. Poongodi, and S. Suganthi

8.1 Introduction

Parkinson's disease (PD) is one of the most common neurodegenerative diseases linked to age, second only to Alzheimer's disease in incidence. At least half a million people are diagnosed as having PD in the United States, and the prevalence of PD is expected to triple in the next 50 years as the average population age increases [1]. The global burden of PD is in a keen increase. The treatment and follow-ups are having a great market. PD is a neurodegenerative disease gradually progressing with no known cause. PD definition does not include neurological symptoms that indicate more severe motor damage or sensory channels that reach beyond the pigmental brain stem nuclei. Such symptoms indicate other neurodegenerative diseases which are also called atypical parkinsonism. These include multiple system atrophy, progressive supranuclear paralysis, striatonigral degeneration, and other conditions that are less common. The term parkinsonism is frequently used in syndromes where etiology is defined, such as stroke-related parkinsonism symptoms, illness, neuroleptic medications, and toxicants.

The European Parkinson's Disease Association (EPDA) has proclaimed April 11, Dr. James Parkinson's birthday, as "World Parkinson's Day" since 1997. Most government agencies and social organizations in several countries chose to conduct the events on April 11 on the theme of PD. Efforts are being made on this day to

R. Sujatha (✉)
School of Information Technology & Engineering, Vellore Institute of Technology, Vellore, Tamil Nadu, India
e-mail: r.sujatha@vit.ac.in

T. Poongodi
School of Computing Science and Engineering, Galgotias University, Delhi, India

S. Suganthi
PG and Research Department of Computer Science, Cauvery College for Women, Tiruchirappalli, Tamil Nadu, India

© The Author(s), under exclusive license to Springer
Nature Switzerland AG 2021
D. J. Hemanth et al. (eds.), *Internet of Medical Things*, Internet of Things,
https://doi.org/10.1007/978-3-030-63937-2_8

increase public awareness of this disorder and encourage early diagnosis, medical care, joint prevention, and management of PD to improve the life of PD patients. As mentioned at the start of this discussion, PD falls under the second common disorder in comparison with the neurodegenerative disorders which are Alzheimer's disease, PD, and motor neuron disorder. Figure 8.1 provides the trend of the common neurodegenerative disorders along with PD in the second place for the past 12 months of duration. Similar to the occurrence of PD, the research perspective is voluminous in nature. The flow of chapter, started with a discussion about PD and its impact followed by the usage of information and communication technologies (ICT) to handle PD; in the perspective of ICT, usage of cloud computing, machine learning, and deep learning to make effective communication and decision-making; glance on biomarker with the home monitoring system for continuous monitoring; and ergonomically suitable wearable devices to substantiate the need of all the stakeholders. Hybridizing with the existing environment to make it highly user-friendly in nature is a great challenge in the initial stages, but effective work in this area made it quite achievable to assist the patient. Followed by that discussed the issues and future direction that meet in the middle between PD and ICT.

8.1.1 Affected by PD

Many studies have shown a more common occurrence of PD in males than females [2]. The probability of people with close contact with PD is found highly prone to illness. Several researches find PD in Whites to be more prevalent than those in

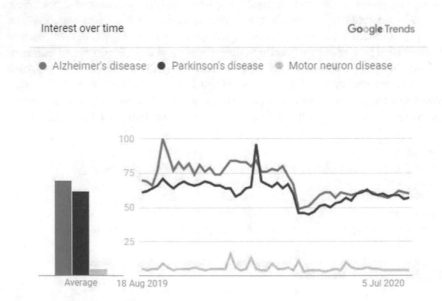

Fig. 8.1 PD is the second common neurodegenerative disorder

African Americans [3]. In comparison with the rural areas, a lifestyle made the urban areas have a higher percentage of PD. Sufferers under the age of 50 have seen a steady rise in recent years due to environmental pollution or toxic chemical substances impacting 1–2% of the population above the age of 65. In PD patients, the most prominent symptoms are hand tremors, stiff or rigid motion, weak balance, and a shuffling gait. Symptoms start slowly and usually get worse over time, leading to chronic illness. According to estimates, at least three million people in China are affected by Parkinson's, and by 2030 the number of patients will rise to five million, representing over 50% of worldwide cases. Early diagnosis is required to ensure the tedious hitting of the patient in both personal and professional life [4–6].

8.1.2   PD Pain

Pain is a typical non-motor symptom of PD, and the severity of pain in patients with PD varies due to the stage of disease, comorbidities, and methods to assess. While many family physicians and also some neurologists can reassure you that PD doesn't hurt, several persons with the disorder are likely to disagree. The nature of pain is heterogeneous and spreads across the whole body. The symptoms, frequency, and pain distribution differ according to sex, race, and stage of the disease. There are different strategies used to overcome the PD pain [7]. Certain signs reflecting this discomfort include fatigue, sleep, and autonomous signs. PD patients' pain can be categorized as:

- Musculoskeletal pain.
- Chronic body pain (central or visceral).
- Fluctuation-related pain, nocturnal pain.
- Orofacial pain.
- Pain with discoloration/edema/swelling.
- Radicular/neuropathic pain.

PD pain is analyzed based on the severity, frequency, and interference. Many tools are available to gauge the same, and they are McGill pain questionnaire, King's PD scale, and the visual analog scale. In this method, the King's is easier, because of the set of questions to the patient to find the severity and occurrence rate. Normal strategy of discussion with the diseased person provides a great chance to know the pain of him [8, 9].

8.1.3   Clinical Symptoms of PD

- Lack of automatic movements of the leg and the arm.
  Patients with PD lose both internal and automatic patterning of the gait. They are afflicted by feasting as they travel and also stopping at the start of ambulation. To

adjust for defective regulation of the gait, PD patients must rely actively on ambulation. They use audio and visual external signals to conquer the feasting. PD patient physiotherapy uses techniques to enable careful cortical walking regulation with consideration to wider steps and complete swing of the arm.

- Freezing.

A normal person when encounter obstacle or dangerous time, they either freeze or escape or avoid but in the case of PD, they get frozen. Decision-making is completely diminished.

- Blindsight.

Missing blindsight elements in patients with PD include decreased sense of motion and impaired grasp saccades. It affects the overall system efficiency [10–12].

8.1.4   Smart Healthcare: PD

Due to the increase in the average living years of the people, it drastically increased the elderly population, and it's required to take care of their health and well-being as the fellow member. To monitor the heart rate, the activity of the brain, breathing rate, blood pressure level, blood sugar level, body temperature, and so on for the person from the remote is the main idea behind the wearable smart devices. Way of integrating the sensors to create the devices with multiple testing ensures the fitness of the device in the healthcare industry. Definitely the importance of this wearable smart device is manifold when a person becomes sick and immovable in nature because of age factor. Based on the criticality, measures generated from the devices and technically called data that flow from the patient's end to the specialist's end serves as the bridge to provide medical support. In the recent decade, many industries provided massive devices to the healthcare industry to monitor the sick from the remote. The performance of the devices and purpose of the devices are being optimized in each phase to ensure quality service to the application users. The affordability, comfortability, and scalability are the key factors to be considered for the best results in any smart healthcare devices. Remotely proving the best service is the success factor that decides the device competency. Digitalization is the buzzword in the general perspective of record maintenance and caretaking of the sick person. In case of PD, the degeneration happens progressively and deteriorates the person's health; indirectly the confidence level declines. Wearable devices in the health market help in gathering the data via the mobile and get transmitted to the central system. As given in Fig. 8.2, the PD patient's raw data reached the system and gets stored in the storage. From there, the neuro-physician constantly monitors from the remote. As the severity increases, it makes the person immobile in nature that pave the way to being completely bedridden in extreme cases. Based on the expertise of the practitioner, the data gets converted into knowledge for providing support in the form of medication to follow for the needy.

8.2   PD Detection

PD is a neurological disease that impairs human speech, motor skills, balance, sensation, emotions, etc. It is a disorder where the general functionalities of neurons are reduced; sometimes it may lead to death or degeneration of neurons. Both invasive (surgeries) and noninvasive (medicine) treatments are commonly followed to manage the severity of this kind of disease.

8.2.1   Cloud-Based PD Detection

Cloud computing is an on-demand networking approach that supports users to access the massive amount of computing resources that are easily available everywhere [13]. The cloud model has specialized features including resource pooling, extensive network accessibility, self-service, measured service, rapid elasticity, etc. It affords services along with the software and hardware required to provide those services [14]. Cloud environment comprises Platform as a Service (PaaS), Infrastructure as a Service (IaaS), and Software as a Service (SaaS). PD can be detected and monitored in the cloud platform by verifying the voice disorders. The classification algorithms such as support vector machine (SVM), random trees, and feedforward backpropagation-based artificial neural network (FBANN) are used for pattern analysis, and voice samples are classified. FBANN obtains the optimal recognition rate by performing offline analysis to diagnose the disease using dysphonia measurement. A patient file is maintained in the cloud database, and the voice sample of the patient is uploaded in the cloud for further analysis. A decision will be

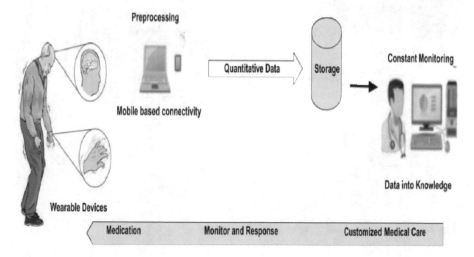

Fig. 8.2 Smart healthcare – PD

made after detecting the voice disorder if the result is positive and the doctor is informed automatically for further treatment. The doctor will download and verify the recorded voice sample, and the decision along with the advice will be uploaded; hence the patient receives information via smartphone. A framework to detect and monitor Parkinson patients in the cloud environment is proposed to facilitate health-care services with minimum resources required. The system would be helpful for the patients in remote areas, places where doctors are not easily accessible, and for the people who are not aware of this disease. Doctors can monitor and detect PD by examining the voice disorder overcloud. Voice samples gathered through phones are adequate to diagnose the disease in spite of their locations.

The patient communicates with the doctor using their smart devices over the cloud. The cloud manager manages data and resource allocation and controls the overall activities that occurred in the cloud platform. A service manager manages the operations performed by virtual machines (VM) which are being used. The media content server handles the voice samples stored in the smart device, sensing server senses the voice input, the transcoding server decodes the voice sample, and streaming server controls the streaming bits. Finally, VM identifies the voice sample based on the voice disorder or dysphonia. In the framework diagram in Fig. 8.3, initially, the patient has to upload the voice sample to recognize whether he has PD or not. The uploaded content will be sent to the cloud system through the cloud manager. Any background noise with the voice recording can disturb the output if not cleared in prior. In the cloud, noises will be removed in the beginning stage itself. In terms of reducing the data volume, downsampling is followed; hence computational complexity and memory allocation are also reduced. Now, noise-free

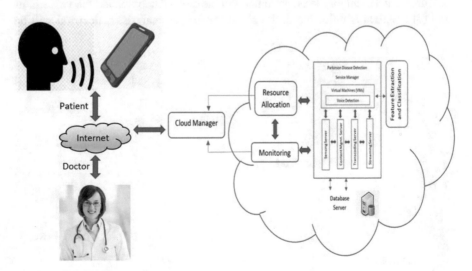

Fig. 8.3 Cloud architectural framework for PD detection

data about the patient is maintained in the cloud database for further processing. For classifying the voice sample, appropriate features are extracted from the speech sample. An optimal feature subset is chosen for classification. The dataset is categorized into training, test, and validation dataset. FBANN is used for classification, and the final output is stored in the cloud database. VM verifies the output having PD or not. Once it is detected, notifications will be sent to the doctor for a diagnosis. Thus, the doctor downloads the voice sample for testing, and the report will be generated accordingly. The same report is communicated with the patient and maintains a timely alert to remind the patient regarding regular monitoring. The voice samples will also be collected for the next testing if the timer is expired.

Patients will be diagnosed whether they are suffering presently or will be suffering after sometime.

Users: Smart devices are embedded with sensors for sensing and recording voices. Healthcare professionals or doctors or PD specialists monitor patients from rural areas.

Cloud manager: Controls and coordinates the overall operations in the framework with the following responsibilities:

(a) User authentication: The cloud manager establishes a connection between the user and the cloud through the Internet. Users can send a request to access the system, and the cloud manager authenticates and registers in the cloud system.
(b) Profile management: The information available and retrieved from the smartphone is maintained in the database.
(c) Communication: The communication is initiated and established between the user and the system.
(d) Contextual information: The contextual information is obtained from the smartphone and stored in the database.

Resource allocation and monitoring: VM is allocated for the particular session along with the web services. Though VM is an interface, it monitors the performance and keeps tracking of resource utilization.

Server responsibilities: Different servers are implemented to accomplish various tasks. The responsibilities of different types of server are described below:

(a) Sensing server: Senses and synchronizes the voice samples retrieved from the patient's smartphone.
(b) Content management server: The complete information is available for the doctors regarding further diagnosis and communication.
(c) Transcoding server: The voice is transcoded and transmitted to the streaming server to proceed with the next step.
(d) Streaming server: Controls the streaming bits obtained in real time.

The step-by-step processes for PD detection in the cloud platform are described in Fig. 8.4.

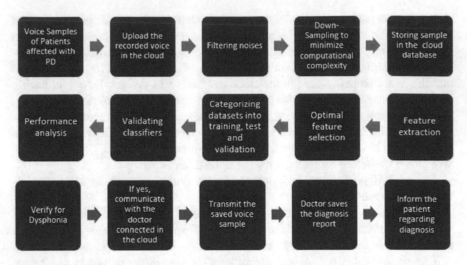

Fig. 8.4 PD Detection in the cloud platform

8.2.2 Machine Learning-Based PD Detection

A gait classification approach in machine learning supports clinicians to identify the several stages of PD. Gait pattern is exploited to examine the human mobility and classifies whether the person is affected or not with PD. The vertical ground reaction force (VGRF) dataset is used to extract the feature based on statistical analysis. The spatial and temporal features are identified based on the correlation. The supervised machine learning algorithms such as support vector machine (SVM), decision tree (DT), Bayes classifier (BC), and ensemble classifier (EC) are used to anticipate the severity level of PD. Most popular methods for diagnosing PD are functional neuroimaging [15], finger motion [16], foot pressure [17], and EMG analysis [18]. The diagnostic procedure in the conventional method depends on the questionnaire and visual observation measurements. Clinicians prepare a questionnaire to investigate the behavioral characteristics, motor symptoms, and daily activities. Based on scaling, the severity level will be found out. In the conventional approach, low efficiency and misclassification are unavoidable because of considering descriptive symptoms as descriptive criteria that fails in affording quantified results [19]. Recently, the paradigm is shifted toward optimal ML solutions that support clinicians to make an early and timely diagnosis by managing the immense volume of clinical data and recognizing the hidden patterns [20]. ML algorithms can diagnose PD using significant factors such as motion signals, speech signals, gait signals, and handwriting dynamics which assist in automatic PD diagnosis in clinical practices.

A noninvasive gait pattern is used to diagnose PD and its severity rating. The spatial and temporal features are extracted mainly from the gait pattern obtained from foot-worn sensors. Initially, the statistical analysis is performed to distinguish between healthy and unhealthy persons. The weight is considered as a significant factor that influences foot pressure. The discriminative feature set is followed to solve the multilevel problems of PD. The ML algorithms include support vector machine (SVM), decision tree (DT), Bayes classifier (BC), and ensemble classifier (EC) which are exploited for classifying different stages of PD. A cross-validation technique is incorporated to improve the classification accuracy and to evade the overfitting problem. The confusion matrix and region of convergence (ROC) curves are used for validating the classifier model. The frequency-domain and time-domain features are used to detect the severity level of PD. These features are not directly communicated with the clinical indicators; instead neurologist support is required to rate the condition of severity level. The gait classification methodology is proposed to predict the severity level of PD. Kinematic and statistical analyses are performed for the gait patterns to extract the biomarkers. The feature set is given as input to the classifier model to predict the severity level of PD. The fluctuation magnitude of the foot-worn sensors is computed to examine the gait pattern for both positive and negative PD patients. The gait classification for PD detection is depicted in Fig. 8.5.

Normally, PD has an impact on the day-to-day walking patterns, and it reduces the step-time, speed, stride interval, and count of footsteps. The gait cycle factors include swing time, stance time, stride time, step time, cadence, stride length, step length, speed, and swing-stance ratio which assist in understanding the disorder of PD. The gait parameters can be considered as spatial and temporal features. Cadence refers to the total count of steps per unit time, and it is the influential factor that shows the larger variability if PD is severe. SVM is more appropriate for regression analysis and binary classification. The major benefit of SVM is that it is capable of handling both linear and nonlinear classification entities. SVM exploits margin and hyperplane as the underlying concept. Moreover, hyperplane in SVM is the linear separator that optimally categorizes the available data into two sets by enhancing the margin size among two classes. Decision tree is a classifier that determines the classification model in a hierarchical structure. The dataset is divided into subsets; each subset is organized in the structure of nodes and edges. Nonlinear relationship is determined iteratively based on the input and output given to the system. Bayes classifier utilizes the probabilistic prediction for classifying the dataset. This algorithm is well appropriate if the input dimensionality size is high and probabilistic information is known. A posterior probabilistic approach is employed to model the relationship between the class variable and feature vectors. Ensemble classifiers' key benefit is that it can combine outputs of several tree classifiers which automatically reduces the risk of poorly constructed classifiers.

8.2.3 Deep Learning-Based PD Detection

Identifying PD is a difficult task that needs the examination of both motor and non-motor symptoms. Gait abnormalities are the significant symptoms that should be considered by the physicians during diagnosis. Gait evaluation is highly challenging that requires expertized clinicians. An intelligent gait analysis approach using a deep learning approach supports physicians in facilitating the diagnostic process.

Fig. 8.5 Gait classification
for PD detection

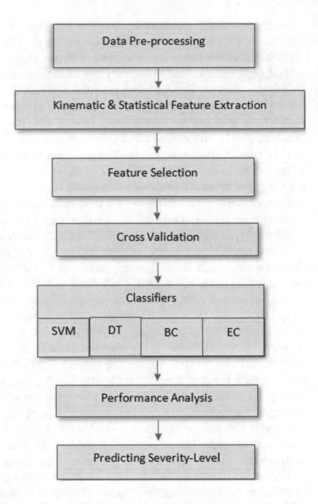

1D convolutional neural network (1D-CNN) is followed to construct a classifier using deep neural network (DNN). 18 1D signals are taken from foot sensors which measure vertical ground reaction force (VGRF). The proposed model predicts the severity of PD using the Unified Parkinson's Disease Rating Scale (UPDRS). The system achieved 98.7% accuracy and 85.3% accuracy in severity prediction. Presently, around ten million people suffered from PD across the world. Unfortunately, there is no proper treatment available to heal from this disorder [21]. Early diagnosis may support and help in improving the patient's treatment. The symptoms considered by the physicians are slowness, shaking, and finding it difficult to walk and initiate the movement even [22]. The widely used tool in PD evaluation is UPDRS, and it comprises 42 questionnaires to cover several perspectives of PD. It covers behavioral characteristics, daily activities, and motor symptoms [23]. The scaling level from 0 to 5 (normal–severe) is followed to detect the severity of a disease. The influence of deep learning in PD is depicted in Fig. 8.6.

Gait abnormalities are significant in diagnosing PD, and it is characterized by a slower gait cycle, shorter and longer swing phases, difference in stride variability, small steps, a flat foot strike, and a shorter swing phase [24, 25]. The features are used by the physicians in the diagnosis process to confirm the disease. However, the evaluation can also have an impact due to factors such as health conditions and age. There is no powerful tool available yet for gait analysis which supports physicians to complete the evaluation process. Some changes in the gait are considered as the symptoms for identifying PD. To identify these features, feature extraction methodologies are commonly utilized. Frequential or temporal tools are also used to distinguish the patterns between the normal and affected gait. Gait is a physiological feature that normally differs for each human according to their intrinsic factors (age, health condition, etc.) Hence, feature extraction and pre-processing are complex and time-consuming as well within the limited capacity. No explicit feature extraction is required in the novel gait classifier which is based on deep learning. There are

Fig. 8.6 Deep learning in PD detection

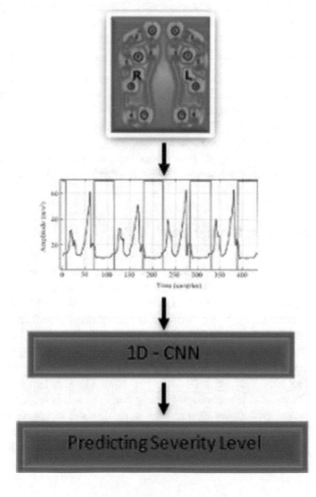

two parts in deep neural network (DNN): i) There are 18 parallel 1D-CNN that processes the VGRF signal received from the foot sensor. The deep features are extracted and concatenated together in each 1D-CNN. ii) In the fully connected network, the concatenated vector is processed to obtain the output in order to make the final decision. The relevant gait features are extracted from several input signals, and the proposed system obtains 98.7 accuracy.

The learning process in deep learning depends on the intermediate layer called the hidden layer. Initially, the input is processed in every layer to obtain a high level of abstraction. In the training phase, accuracy is increased using optimizers that automatically reduce the loss in each iteration. Thus, with adequate data, features are extracted from the given input. The neurons in each layer calculate the summation of all the inputs that are connected together [26]. However, a nonlinear activation function is implemented to yield the neuron output. The learning is completely based on the weights W that are assigned. The optimal value of W is obtained in order to find out the nearest estimate of the real value. The regularization technique, named dropout, is generally used to avoid the over-fitting problem. It minimizes the over-fitting by closing some neurons that are randomly chosen in each iteration [27]. Some layers include convolutional, max-pooling, and fully connected layers which are used in the proposed model. They are described briefly:

- Convolutional: A spatial convolution is accomplished among the inputs and the filters. Generally, the filters will have weights that are considered as the learned elements in each layer.
- Max-pooling: The input is sampled to yield an output with a lesser dimension by choosing the maximum value.
- Fully connected: Associates the outputs obtained from convolution layers to form the final output. Hence, the decision is made at a higher level.

8.3   PD: Biomarkers

PD is a chronic neurodegenerative condition often caused by a lack of brain dopamine. Dopamine is the neurotransmitter that is responsible for several functions like movement, inspiration, and so on. In the case of PD, the level of the chemical is decreased due to the death of dopaminergic cells. It leads to a deficiency in motor and deficit of cognitive nature. Diagnosing at the earlier stage is not possible and reveals when a large number of cells died [28]. Finding reliable molecular biomarkers is the required research area that helps in tracking and monitoring PD in a perfect way. Study of brain examples from a huge selection of PD clients unveiled that the method is pathological fairly consistent. It is important to consider when vagal nuclei within the medulla are compromised by some pathological process, long tract signs are often related. A pure vagal nerve or more frequently, involvement of a vagal nerve suggest a pathology outside the medulla. In stage 2, dorsally deteriorating lesions in the engine, inclusions grow into the reduced raphe nuclei, and in the locus coeruleus, Lewy neurites could be observed. In phase 3, the SN is impacted. In phase 4, lesions come in the cortex, especially within the mesocortex that is temporal. The disease tends to be in the adjoining temporal neocortical industries in phase 5, while cortical involvement is demonstrably seen in phase 6. Notably, intellectual status correlates with all the neuropathological phase [29, 30].

8.3.1   Need for Biomarkers

Biomarkers have a diverse and effective approach to understanding the neurological disease continuum, with applications in observational and comparative epidemiology and clinical trials followed by screening with diagnosis [29, 31, 32].
Abilities of biomarkers:

- Establishment of impact and variability modification.
- Improved group and specific danger assessments.
- Delineation of activities between condition and visibility.
- Establishment of dose-response.
- Lowering of misclassification of exposures or danger facets and infection.
- Recognition of mechanisms through which infection and publicity are associated.
- Recognition of very early occasions into the history that is normal.

8.4   Home Monitoring System for PD

PD is a common neurodegenerative disorder caused due to the degeneration of neurons signalized by both motor and non-motor symptoms which affects the quality of life of the person being affected. The motor symptoms include tremor, rigidity, slowness of movements, muscular cramps, and postural imbalance. In addition, non-motor symptoms such as depression, dementia, hallucinations, sleep disturbances, etc. are also associated with the disease. It is studied that non-motor symptoms can be observed 5–7 years before the onset of motor complications and are related to the neuropathological changes in the brain. PD symptoms should be monitored in a reliable and accurate way for the early diagnosis and better treatment thereby enhancing the quality of life. The conventional subjective assessment of the disease is very tedious, inaccurate, and time-consuming. The patients normally visit clinics on a regular basis for tracking the progression of the disease after the onset of the disease. This is very tedious for PD patients with mobility problems. Also, the fluctuations in the motor symptoms which may depend upon a particular environment cannot be detected in the short time of the visit and that too in a clinical setup. Hence, a long-term monitoring of the patient with an objective assessment of the

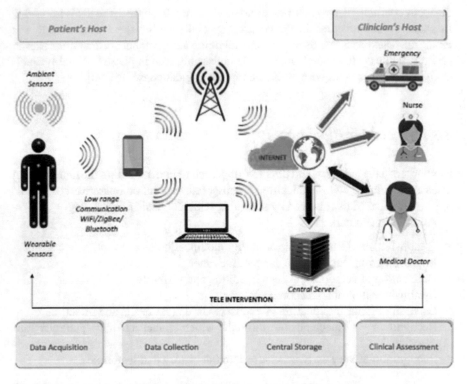

Fig. 8.7 A general design of Home Monitoring System used in PD

disease providing accurate measurements is needed for the healthcare professionals to get better insights into the disease.

Home monitoring plays an effective role in the early detection of disease by continuous monitoring and collecting sensor data. The home monitoring system adapted for such a purpose consists of wireless wearable sensors attached to the patient and the data being sent to a remote clinical site through a web-based system. With advancements in sensor technology, and wireless web-based applications, it is feasible to monitor a patient's health at home continuously and observe various stages of development of the disease starting from the early symptoms to disease progression and later management of the disease with value-added healthcare services. The home monitoring system plays an efficient and cost-effective system for enhancing the standards of healthcare systems. A general design of the home monitoring system used in PD is given in Fig. 8.7.

8.4.1   Wearable Sensors

Wearable sensors are sensors integrated into small devices that are worn by humans or directly implanted in the body which are used in health monitoring and generating health-related data. Wearable technology designed for each application is different with respect to usage, performance, and environment. Few wearables include smart-watches, wrist bands, smart jewelry, smart jackets, smart glasses, skin patches, etc. They are characterized by low cost, accurate measurements, easy usage, low power consumption, and less weight and are a useful tool in determining various symptoms associated with PD. Usually, wearables are composed of a minimum one sensor, a signal processor for processing the sensed raw signals, and a display to reveal the physiological parameters of the user. It consists of essential sensors incorporated along with analysis and data management software for the quantification of the data collected. The wearables have the capability of sensing motion and various parameters related to physiological and biochemical processes with advancements in fabrication and packaging technologies [33]. There are noninvasive sensors integrated in wearables that make use of optic-sensing, oscillometric, and infra-red sensing [34]. The measurement duration of capturing the data depends on the battery and the amount of memory. The sampling rate of the data depends upon the physiological signal captured. Home monitoring involves wearable sensors that play a significant role in the automatic data collection, diagnosis, treatment, and management of PD with less manual interaction. For PD they are applied in the areas of early diagnosis, analyzing motor performances and fluctuations, tremor detection, and home monitoring [35]. They are prominently used in the detection of motor symptoms [36]. For patients with PD, the motor symptoms gradually worsen with time, and they have to be monitored and recorded for quantification of the data to provide personalized treatments. The following are a few wearable sensors used in the applications of PD:

8.4.1.1   Accelerometer

Accelerometers are used in the home monitoring of PD patients who are characterized by postural imbalance and abnormal gait. It is used to measure linear non-gravitational acceleration which when integrated with a patient's body produces voltage readings depending on the vibrations created due to movements.

8.4.1.2   Gyroscope

Gyroscope is used to determine the orientation of the object with respect to Earth's gravity. They can determine the angular position and the rate of rotation along a particular axis, thereby detecting the orientation of the object. The movements and position of the body can be determined using this.

8.4.1.3 Magnetometer

Magnetometer helps to find the orientation of an object by measuring the earth's magnetic field. Inertial measurement unit (IMU) is an electronic device that has an accelerometer; gyroscope, and sometimes magnetometer packed together which is a commonly used wearable motion sensor.

8.4.1.4 Speech Recognition Sensor

There is a strong proof that speech disorders have a close relation to PD and is one of the initial symptoms identified in PD. The problems associated with speech disorder include a reduction in the volume of voice, difficulty in uttering words, and inability to produce normal vocal sounds. Numbers of speech signal algorithms have been proposed for the detection of the severity of PD symptoms. Vowel phonation and running speech are used to assess the severity of speech impairment. Running speech provides more convincing results but is more complicated. Thus vowel phonation is more generally used in speech assessment. The patient speaks into the microphone, and the speech is recorded into a telemonitoring device placed at the patient's home. Vowel phonation is conducted by requesting the patient to make recordings by withstanding the vowel sound as long as possible. Later the recordings are examined by using speech signal algorithms. Statistical mapping of the algorithm is done with the UPDRS, and the results are predicted by the medical examiner.

8.4.1.5 Electromyography (EMG) Sensor

Electromyography uses an EMG sensor that can be used in the detection of neuromuscular abnormalities by measuring the electrical activity of muscles. It is used to detect abnormalities such as nerve dysfunction and muscle dysfunction.

8.4.1.6 Electroencephalogram (EEG) Sensor

EEG scan done using EEG sensor is a noninvasive method which is used to measure the electrical activity of the cerebral cortex by placing multiple electrodes on the scalp. It is used in the identification and analysis of various neurological disorders by recording the brain's activity. Non-motor symptoms associated with PD such as sleep disorders, depression, epilepsy, etc. can be detected. The signals collected by the EEG sensor are amplified and digitally stored in a computer or a handheld device for later processing.

8.4.2   Smart Phone Applications

The wide use of smartphones by most of the people and the increasing adoption of information technology enable value-added service being provided with increased interaction among the patients and the healthcare providers. Smartphones have built-in voice recorders, speakers, accelerometers, gyroscope, proximity sensors, and global positioning systems (GPS) which aids in the healthcare process by monitoring the movements and activities of the patients. Home monitoring through smart applications consists of wearable sensors linked to smartphone applications via wireless communication or embedded sensors which are responsible for sensor data collection, cloud servers acting as data storage, and client applications for user interaction. Mobile applications in smartphones can be divided as applications that are specifically designed for PD and the applications that may be useful for PD [37]. Applications that are specifically designed include information apps (provides information related to PD for the patients and the healthcare providers), assessment apps (analyze various parameters related to PD by testing the patients), and treatment apps (provides guidelines for therapies and treatments related to PD). Applications useful for PD help only in the management of the disease and do not provide services specific to PD. Some mobile applications require additional sensors or devices for complete functioning. Smartphone applications are an important tool in the accurate diagnosis of the early stages of PD [38]. Also, the smartphone interface should be designed in such a way that it supports PD patients who are badly impaired with motor functions. The health apps should satisfy the requirements of accessibility, safety, privacy, security, portability, stability, trustability, certifiability, and usability. But unfortunately, most of the apps do not comply with all the requirements or the required standards.

8.4.3   Hybrid of Web-Based Applications with Wearable Devices

The web-based system integrated with wearable devices can be used in the home monitoring of PD patients. It involves data being collected by the wearable sensors which are being sent to a handheld device or an access point through wireless protocols and transmitted to a remote site through the Internet for further analysis and tracking of the disease. By using a hybrid of web-based with wearable devices, an electronic diary of motor symptoms can be maintained. Home monitoring of PD patients with the hybrid of wearable devices and web-based applications consists of three components which include the patient's host, the central portal server, and the health provider's host. The patient host consists of the necessary wearable sensors worn on the patient's body connected together through a wireless technology. They are located in the patient's home and are responsible for collecting sensor data and transmitting it to the central portal server. The central portal server provides a reliable and secure data storage and is responsible for coordinating real-time data collection. And the health provider's host is responsible for monitoring the patient and communicates with him through the interface provided on both the ends.

8.4.4 Hybrid of Ambient Sensors with Wearable Devices

Ambient sensors are placed at fixed locations inside the patient's house to monitor their movements. There are different kinds of ambient sensors such as temperature sensors, sound sensors, pressure sensors, etc. which capture the changes in the environment. Ambient sensors also reduce the burden of wearing more wearables on the body, thereby increasing the comfort of the patients. The sensor data from all the ambient sensors are received by a base station which then sends it to a server. The communication between the base station and all other ambient sensors takes place with the help of a protocol (e.g., ZigBee protocol). The data obtained through the ambient sensors are useful in determining the patterns and volume of movement by the PD patients which provide added accurate information when combined with the sensor data collected through the wearables. The machine learning techniques when applied to the data collected from multi-modal sensors provide accurate information in determining the time and the pattern of activities done by the patients.

8.5 Open Research Challenges

8.5.1 Privacy and Security

Wearables are meant for continuous data collection generating huge volumes of personal health data. The security and privacy of these data have to be preserved by implementing authenticating protocols and encryption methods which is a major technical challenge. But wearables have constrained battery backup, computing power, and memory which makes it unsuitable for implementing heavy security protocols. Also, the transfer of long keys is difficult as the payload size of the protocols (Bluetooth LE) is limited. So low power encryption schemes with efficient and secure key sharing methods should be adopted.

8.5.2 Energy Efficiency

Wearables are small compact and lightweight devices that constraint their battery capacity. Making of energy-efficient batteries with efficient sensing and energy harvesting methods is another challenge. Efficient sensing in batteries can be improved by adopting various sensor selection methods by which a subset of sensors are turned on while keeping other sensors off depending upon the need to reduce power consumption. Also, sensors that use less energy (e.g., inertial sensors) can be used in place of high energy-consuming sensors. Adaptive sampling is another technique which minimizes power consumption by reducing the number of samples taken during a particular interval of time. However, all these methods should be adopted without compensating for the performance of the system and the precision of the data collected. The wearables can be self-powered by using mechanical, solar, and thermal energy by using energy harvesting (EH) technologies [39]. But determining the type of harvesting method is dependent on the pattern of activities carried by an individual user.

8.5.3   Packaging Issues

Wearable is an integration of various components such as sensors, computer chips, speakers, cameras, etc. which have to be packaged into a small space in the right form for efficient functioning. Packaging all the components in a compact space is a big challenge. Nowadays molded interconnect devices (MIDs) are being used in which the components like antenna can be placed in the house itself. Also as wearables are naturally exposed to dust, liquid, water, etc., they should be sealed properly coping with the standards of IP rating (Ingress Protection Rating or International Protection Rating).

8.5.4   Durability and Robustness

The durability and robustness of the device should be ensured as it is crucial for continuous data generation. The wearables are prone to operate under varying environments and under different activities performed by the patients. They should be designed in a smart way that it can function even if there is any misalignment with the body. The various parts including the battery should be designed in a robust way to withstand varying situations.

8.5.5   Communication with the Internet

Most of the wearables are usually paired with other handheld devices and cannot connect directly with the Internet due to lack of chipsets which supports long-range communications. They mostly use short-range communication protocols such as Wi-Fi and Bluetooth. Also, the wearable operating system is not designed to support such direct communication. But very few protocols that support long-range communications in wearables are coming up.

8.5.6   Design Issues

8.5.6.1   Personalized Design

Every individual is distinct and has different factors affecting their health, and so the analysis and treatment process has to be followed in a personalized way for attaining the best results. So wearable devices calibrated in a personal way depending upon the individual need will help in accurate analysis and treatments. The machine learning approaches should be tailored to handle data that are generated for a particular individual instead of using big data for making major treatment plans. Thus personalized calibration of devices is another major challenge and incurs more cost.

8.5.6.2   Interface Design

PD patients are impaired physically and cognitively, and so wearables, smartphone applications, and ambient assisted living (AAL) should be designed ensuring zero effort from the users. The interfaces if used should be very simple and easy for the patients to use. The caretakers should be considered in the designing ensuring easy usage of the system and thereby providing proper care.

8.5.7   Social Issues

People are naturally reluctant to accept the new systems and learn to use any new technology. Importance should be given to design systems with a simple user interface. Also, proper training and support should be provided to achieve maximum utilization of the system.

8.5.8   Legal Issues and Ethical Issues

Privacy and security of the data should be ensured by well-established rules and regulations. Also, user rights should be protected by adopting proper reimbursement procedures and in case of any misuse.

8.6   Conclusion and Future Direction

Number of applications under research to serve the purpose of taking care of PD patients are evolving. Extensive research is required in this field and scope also available because the type of symptoms is so bothering. Computer-aided diagnosing is required to be optimized. PD is not curable on the whole but is required to be tracked to provide a comfortable space. During this time of COVID pandemic, it's mentioned to take extra care of the PD patients. The possibility of attack is more because they already have breathing difficulty and antibodies are not so strong to withstand the issues. PD cause is uncertain to a great extent, so finding at the early stage and starting medication is the feasible one. Technology join hands to provide optimal comfort.

References

1. Goldman, S. M., & Tanner, C. (1998). Etiology of Parkinson's disease. In J. Jankovic & E. Tolosa (Eds.), *Parkinson's disease and movement disorders* (3rd ed., pp. 133–158). Baltimore: Williams and Wilkins.
2. Haaxma, C. A., Bloem, B. R., Borm, G. F., Oyen, W. J. G., Leenders, K. L., Eshuis, S., Booij, J., Dluzen, D. E., & Horstink, M. W. I. M. (2007). Gender differences in Parkinson's disease. *Journal of Neurology, Neurosurgery & Psychiatry, 78*(8), 819–824.
3. Bailey, M., Anderson, S., & Hall, D. A. (2020). Parkinson's disease in African Americans: A review of the current literature. *Journal of Parkinson's Disease*, 831–841.
4. Kim, K. S. (2017). Toward neuroprotective treatments of Parkinson's disease. *Proceedings of the National Academy of Sciences of the United States of America, 114*(15), 3795–3797.
5. Zhang, Z. X., Roman, G. C., Hong, Z., Wu, C. B., Qu, Q. M., Huang, J. B., et al. (2005). Parkinson's disease in China: prevalence in Beijing, Xian, and Shanghai. *Lancet, 365*(9459), 595–597.
6. Dorsey, E. R., Constantinescu, R., Thompson, J. P., Biglan, K. M., Holloway, R. G., Kieburtz, K., et al. (2007). Projected number of people with Parkinson disease in the most populous nations, 2005 through 2030. *Neurology, 68*(5), 384.
7. Tai, Y. C., & Lin, C. H. (2020). An overview of pain in Parkinson's disease. *Clinical Parkinsonism & Related Disorders, 2*, 1–8.
8. Chaudhuri, K. R., Rizos, A., Trenkwalder, C., Rascol, O., Pal, S., Martino, D., et al. (2015). King's Parkinson's disease pain scale, the first scale for pain in PD: An international validation. *Movement Disorders, 30*(12), 1623–1631.
9. Skogar, O., & Lokk, J. (2016). Pain management in patients with Parkinson's disease: Challenges and solutions. *Journal of Multidisciplinary Healthcare, 9*, 469.
10. Diederich, N. J., Uchihara, T., Grillner, S., & Goetz, C. G. (2020). The evolution-driven signature of Parkinson's disease. *Trends in Neurosciences*.
11. Mirelman, A., Bernad-Elazari, H., Thaler, A., Giladi-Yacobi, E., Gurevich, T., Gana-Weisz, M., et al. (2016). Arm swing as a potential new prodromal marker of Parkinson's disease. *Movement Disorders, 31*(10), 1527–1534.
12. Ha, A. D., & Jankovic, J. (2012). Pain in Parkinson's disease. *Movement Disorders, 27*(4), 485–491.
13. Mell, P., & Grance, T. (2011). *The NIST definition of cloud computing* (NIST special publication 800–145). National Institute of Standards and Technology NIST.
14. Armbrust, M., Fox, A., Griffith, R., Joseph, A. D., Katz, R. H., Konwinski, A., Lee, G., Patterson, D. A., Rabkin, A., Stoica, I., & Zaharia, M. (2009). Above the clouds: A Berkeley view of cloud computing. *February, 10*.
15. Appel-Cresswell, S., Doudet, D., Sossi, V., et al. (2011). Functional neuroimaging in Parkinson's disease. *Expert Opinion on Medical Diagnostics, 5*(2), 109–120.
16. Kim, B., Lee, W. W., Kim, A., Lee, H. J., Park, H. Y., Jeon, H. S., Kim, S. K., Jeon, B., & Park, K. S. (2018). Wrist sensor-based tremor severity quantification in Parkinson's disease using convolutional neural network. *Computers in Biology and Medicine, 95*, 140–146.
17. El Maachi, I., Bilodeau, G.-A., & Bouachir, W. (2020). Deep 1d-convnet for accurate Parkinson disease detection and severity prediction from gait. *Expert Systems with Applications, 143*, 113075.

18. Robichaud, J. A., Pfann, K. D., Leurgans, S., Vaillancourt, D. E., Comella, C. L., & Corcos, D. M. (2009). Variability of EMG patterns: A potential neurophysiological marker of Parkinson's disease? *Clinical Neurophysiology, 120*(2), 390–397.
19. Zhao, A., Qi, L., Li, J., Dong, J., & Yu, H. (2018). A hybrid spatio-temporal model for detection and severity rating of Parkinson's disease from gait data. *Neurocomputing, 315*, 1–8.
20. Belić, M., Bobić, V., Badža, M., Šolaja, N., Đurić-Jovičić, M., & Kostić, V. S. (2019). Artificial intelligence for assisting diagnostics and assessment of Parkinson's disease–a review. *Clinical Neurology and Neurosurgery, 105442*.
21. National Health Service: https://www.nhs.uk/conditions/parkinsons-disease/treatment/. Accessed 26 Sept 2020.
22. Jankovic, J. (2008). Parkinson's disease: Clinical features and diagnosis. *Journal of Neurology, Neurosurgery & Psychiatry, 79*(4), 368–376.
23. Fahn, S., Elton, R., Marsden, C. D., Calne, D., & Goldstein, M. (1987). The unified Parkison disease rating scale. *Recent Developments in Parkinson's Disease, 2*, 153–163.
24. Pistacchi, M., Gioulis, M., Sanson, F., De Giovannini, E., Filippi, G., Rossetto, F., et al. (2017). Gait analysis and clinical correlations in early Parkinson's disease. *Functional Neurology, 32*(1), 28.
25. Reich, S. G., & Savitt, J. M. (2019). Parkinson's disease. Medical Clinics of North America. *Neurology for the Non-Neurologist, 103*(2), 337–350. https://doi.org/10.1016/j.mcna.2018.10.014.
26. Poongodi, T., Sumathi, D., Suresh, P., & Balusamy, B. (2020). *Deep learning techniques for Electronic Health Record (EHR) Analysis* (Studies in computational intelligence) (Vol. 903, pp. 73–103). Springer.
27. El Maachi, I., Bilodeau, G.-A., & Bouachir, W. (2020). Deep 1D-Convnet for accurate Parkinson disease detection and severity prediction from gait. *Expert Systems with Applications, Elsevier, 143*, 1–7.
28. Levy, O. A., Malagelada, C., & Greene, L. A. (2009). Cell death pathways in Parkinson's disease: Proximal triggers, distal effectors, and final steps. *Apoptosis: An International Journal on Programmed Cell Death, 14*(4), 478–500. https://doi.org/10.1007/s10495-008-0309-3.
29. Emamzadeh, F. N., & Surguchov, A. (2018). Parkinson's disease: Biomarkers, treatment, and risk factors. *Frontiers in Neuroscience, 12*, 612.
30. He, R., Yan, X., Guo, J., Xu, Q., Tang, B., & Sun, Q. (2018). Recent advances in biomarkers for Parkinson's disease. *Frontiers in Aging Neuroscience, 10*, 305.
31. Mayeux, R. (2004). Biomarkers: Potential uses and limitations. *NeuroRx, 1*(2), 182–188.
32. Chen-Plotkin, A. S., Albin, R., Alcalay, R., Babcock, D., Bajaj, V., Bowman, D., et al. (2018). Finding useful biomarkers for Parkinson's disease. *Science Translational Medicine, 10*(454).
33. Koydemir, H. C., & Ozcan, A. (2018). Wearable and implantable sensors for biomedical applications. *Annual Review of Analytical Chemistry, 11*, 127–146.
34. Rashidi, P., & Mihailidis, A. (2012). A survey on ambient-assisted living tools for older adults. *IEEE Journal of Biomedical and Health Informatics, 17*(3), 579–590.
35. Rovini, E., Maremmani, C., & Cavallo, F. (2017). How wearable sensors can support Parkinson's disease diagnosis and treatment: A systematic review. *Frontiers in Neuroscience, 11*, 555.
36. Botros, A., Schütz, N., Camenzind, M., Urwyler, P., Bolliger, D., Vanbellingen, T., et al. (2019). Long-term home-monitoring sensor technology in patients with Parkinson's disease—Acceptance and adherence. *Sensors, 19*(23), 5169.
37. Reya, M. L.-d., Vela-Desojob, L., & Cano-de la Cuerdaa, C. R. (2019). Mobile phone applications in Parkinson's disease: A systematic review. *Science Direct, 34*(1), 38–54.
38. Son, H., Park, W. S., & Kim, H. (2018). Mobility monitoring using smart technologies for Parkinson's disease in free-living environment. *Collegian, 25*(5), 549–560.
39. Seneviratne, S., Hu, Y., Nguyen, T., Lan, G., Khalifa, S., Thilakarathna, K., … & Seneviratne, A. A survey of wearable devices and challenges. IEEE Communications Surveys & Tutorials, 19(4), 2573–2620, (2017).

Chapter 9
Machine Learning and Internet of Things Techniques to Assist the Type I Diabetic Patients to Predict the Regular Optimal Insulin Dosage

T. Jemima Jebaseeli, D. Jasmine David, and V. Jegathesan

9.1 Introduction

Every cell in the body cannot produce energy directly. The consumed food is digested through the pancreas. It produces the insulin hormone for a healthy person after a meal and digests the food to produce glucose. The beta cells of the pancreas create insulin. Also, insulin helps to use the produced glucose and convert into energy by carrying through the bloodstreams to the muscle, liver, etc. In many people, insulin is not produced adequately or sometimes inadequately. The people have to take insulin medication [1].

Consuming insulin helps to manage sugar levels. It is injected through a pen or pump [2]. The insulin pen is disposable or reusable. The disposable pen contains insulin dosages. For a reusable pen, the cartridge needs to be refilled. The insulin pump will be operated by a battery. The insulin pump is a programmable device to supply insulin under the fatty abdomen in the body. It has to be worn under the skin through a thin plastic tube called infusion set or giving set. Also, insulin is given through an inhaling method called Afrezza. This will not be given to people having

T. Jemima Jebaseeli (✉)
Department of Computer Science and Engineering, Karunya Institute of Technology and Sciences, Coimbatore, Tamilnadu, India
e-mail: jemima_jeba@karunya.edu

D. Jasmine David
Department of Electronics and Communication Engineering, Karunya Institute of Technology and Sciences, Coimbatore, Tamilnadu, India
e-mail: jasmine@karunya.edu

V. Jegathesan
Department of Electrical and Electronics Engineering, Karunya Institute of Technology and Sciences, Coimbatore, Tamilnadu, India
e-mail: jegathesan@karunya.edu

© The Author(s), under exclusive license to Springer
Nature Switzerland AG 2021
D. J. Hemanth et al. (eds.), *Internet of Medical Things*, Internet of Things,
https://doi.org/10.1007/978-3-030-63937-2_9

breathing difficulties. The jet injector sprays the insulin without pricking the needle. The injection port will be placed below the skin to supply for 24 h. There is no insulin available in the form of a pill. Scientists are trying for 80 years to produce it. But the research alternatives are still going on, because insulin is a hormone to flow through the bloodstream. When it is taken through the oral method, it reaches the digestive system, where it is a protein-based hormone, and it will be broken; hence there is no use of it. In the present system, insulin dosages are recorded manually by the physicians for treatment [37].

Type I diabetes occurs among children and adults under 30 years. It is known as insulin-dependent diabetes mellitus (IDDM). Type II diabetes occurs among all adults. It is known as non-insulin-dependent diabetes mellitus (NIDDM). Gestational diabetes occurs among pregnant women during the pregnancy period. Prediabetes occurs among people from 18 years onward. It is just before the stage of type II diabetes. The bloodstream glucose level will be high due to insulin resistance, and beta cells do not produce enough insulin. If a person is of type I diabetes, then they must have to take insulin regularly. If they fail to take insulin, then it leads to a life-threatening disease called diabetic ketoacidosis (DKA) [2]. It occurs due to lack of insulin, and the body started to use fat instead of glucose for energy. This releases ketone chemicals. The ketones start to make the blood more acidic, called acidosis. This can be checked through blood and urine test. Type II diabetic patients are also advised to take insulin. If they failed to take, they may become ill due to high blood sugar. Hence insulin maintains the body from too high glucose level of hyperglycemia or too low glucose level of hypoglycemia.

As shown in Table 9.1, there are five types of insulin to manage diabetes: rapid-acting insulin, short-acting insulin, mixed insulin, intermediate-acting insulin, and long-acting insulin [3]. The types of insulin are described with the speed at which it should dissolve in the blood and its effects work.

Insulin causes the following side effects such as hypos. This happens when a person consumes more insulin. The general side effects cause nausea or flu kind of symptoms within 72 h [4]. Bruises, bleeding, itching, skin irritations, and lumps happen in the skin due to injecting the needle in the same place or not changing the needle after usage. Weight gain happens at the starting stage. There are three types of insulin depending on its source of production. Human insulin is made from laboratories. Analogue insulin is produced by changing the insulin molecule from strings of beads through genetic engineering. Animal insulin is produced from cow or pig that suits only a few people [5, 6].

Table 9.1 Types of insulin and working behavior [3]

Types of insulin	Onset after injection	Peaks	How long it lasts
Rapid acting	15 min	1 h	2–4 h
Short acting	30 min	2–3 h	3–6 h
Intermediate acting	2–4 h	4–12 h	12–18 h
Long acting	Several hrs	Does not	24 h

There is a chance for 578 million people who may have diabetes in 2030. Diabetes is a threat to the nations because it impacts the global health of people. It influences the economy of the country. Still many countries do not have the plan to manage diabetes. There are half a billion people with diabetes worldwide, and it is an alarming situation. In 2019, the estimation is 463 million and reaches to 578 million by 2030 and 700 million by 2045. The following are statistics projected by IDF (International Diabetes Federation) about diabetic populations [6, 7]:

- 1 in 11 grown-ups (20–79) years have diabetes (463 million) population.
- 1 in 2 grown-ups with diabetes is undiscovered (232 million) population.
- 1 in 5 individuals with diabetes is over 65 years old (136 million) population.
- 10% of worldwide well-being use is spent on diabetes (USD 760 billion).
- 1 in 6 live births (20 million) is with hyperglycemia during pregnancy where 84% are having gestational diabetes.
- 3 in 5 (79%) of individuals with diabetes live in low- and middle-income nations.
- Over 1.1 million youngsters and grown-ups are under 20 years with type I diabetes.
- 1 in 13 grown-ups (20–79) years have impeded glucose resilience (374 million) individuals.
- 2 in 3 individuals with diabetes live in urban territories (310.3 million) populace.

Bariatric surgery is an alternative treatment of diabetes by surgery. It regulates metabolism, artificial pancreas, transplantation of the pancreatic islet and replaces the pancreatic islet cells affected [5, 6]. Reducing the weight, good diet, and exercises may help most of the people to go to reverse the insulin resistance.

The organization of the contents is as follows. Section 9.2 describes the literature survey of various researchers in this domain. Section 9.3 presents the dataset considered for the proposed research. Section 9.4 elaborates on the proposed methodology to identify the dosage of insulin required for type I diabetic patients based on their glucose level in the human body. Section 9.5 discusses the experimental results and comparison with existing techniques. At last Sect. 9.6 concludes the proposed work and the enhancement of the work in the future aspect.

9.2 Literature Review

Machine learning is one of the major advancements in computing technology. It provides an automatic analysis of data based on the input. The network is trained by supervised learning. But in clinical analysis, the unsupervised training models are used for prediction and monitoring. Thus the machine learning-based algorithms are widely considered for a lot of medical diagnoses. There are various techniques proposed by researchers in the field of predicting the insulin dosage of the insulin pump for the automatic supply of insulin required at every second in the insulin patient. Despite all the existing algorithms, the proposed algorithm considers the reliability of the system and enhances the accuracy of the overall performance to predict the insulin dosage. The various research work done by the researchers in this domain is presented as follows. The identified research gaps are solved through the proposed technique.

The insulin pump is used to provide insulin to type I diabetic patients from the 1970s. This is provided through the computer-aided system. But the transmission of data through wireless communication is not encrypted, so there are chances for data loss and leakage. So it is vulnerable to use due to the introducers. The access to the device may be hacked by someone to change the insulin dosage or stopping its usage [8]. The usage setting has to be modified by the patients [9]. The application is not privileged to store the transformed glucose in some system which was coded using JAVA [10].

Heena et al. [11] proposed a Bayesian network model to find the glucose infusions in the human body and also find the false probability of insulin dosage sent through the wireless medium. The features considered for training the classifiers are the records of patients, time, and glucose measurement value throughout the day. Classifier labels 0 and 1 are used to label the output as regular and unusual insulin measurement. But the system fully depends on the security of all other connected device links [30]. There is a probability of attackers in the DLRT system at the wireless ink 1, 2, 3, and 6. It achieved the reliability of 18% when the connected links are not secured [12]. The smart city depends on the population, so this has to be taken care of by the government by developing efficient techniques.

Inadequate control of diabetes leads to retinopathy, maculopathy, neuropathy, nephropathy, and macrovascular diseases. To predict diabetes, hemoglobin (HbA$_{1c}$) in the bloodstream is tested through the laboratory. HbA$_{1c}$ test gives the details of the mean value of glycemia from 8 to 12 weeks [13]. If HbA$_{1c}$ is below the range of 5.7% that is normal, 5.7%–6.5% is the indication of prediabetes, and ≥6.5% is diagnostic of diabetes. The recommended target of below 7.0% or > 6.5% is possible through proper medications.

Metformin (MET), sitagliptin (STG), and saxagliptin (SAG) are oral diabetic medicine for patients [14]. All these combinations are available as a single zipmet tablet. Dennis et al. [15] proposed a metformin therapy for diabetic patients. The system considers only supplying metformin to the patients through machine learning algorithms [31, 32, 36, 38–40]. The system achieved an overall AUC of 0.58 to 0.75. Comparatively, it is very less accurate with the other recent approaches. The problem in the system is failed to consider the primary and secondary metformin details of the patients. At least a year of consumption must be considered for analysis. Another thought is that patients were most certainly not screened for earlier treatment with glucose bringing down other than metformin [33]. Therefore, at that point, some may have started treatment with another prescription with metformin escalation. Be that as it may, in clinical practice, most patients start with metformin as the first-line specialist and are then heightened with different drugs rather than the other way around. Moreover, the examination was intended to reflect clinical practice however much as could be expected, whereby authentic treatment information may not be promptly accessible. The reinforcement learning algorithm [16] has been used to provide insulin dosage to an existing diabetic patient while coming for cancer treatment. The sampling time considered in this method is only for 5 days.

Ning et al. [17] proposed the incremental learning and echo state network (ESN) to predict the blood sugar. The ESN is designed with few parameters. This method will not be managed if the patient suffered due to unforeseen disruptions in the case of conception of significant insulin dose or high intensity of exercises.

Therese et al. [18] introduced the clinical andon board (CAB) alert system to indicate the dysglycemia state of the people. It considers the people pre/post insulin details for analysis purposes. The algorithm is not filling the gap of finding out the gap between insulin insufficiency and insulin resistance physiology.

The idea proposed by Shoaze et al. [19] is to find the risk of readmission of patients in a hospital at critical times. Nowadays readmission in a hospital is a risk, and it is costly and also limited. Avoiding readmission of patient consistency improves their life as well as the economy of the healthcare system. The readmissions are reduced through analyzing the patient's healthiness through phone calls and follow-ups up to 30 days. This is considered as pre-discharge intervention, post-discharge interventions, and the transitional interventions, because of interventions of the patient that greatly improve the health to avoid risks [28, 29]. The major reasons for readmission are identified by the researchers which are diabetes. It is identified through the random forest algorithm. The occurrences of readmission through diabetic patients are managed through insulin pumps.

Every patient has different reactions to different insulin consumption. Also, every person has different blood glucose characteristics in their body by nature. D1NAMO dataset was used to study the BG level, basal, and bolus insulin of four patients. The psychological modal improved bolus insulin intake. The ensemble method is used for insulin prediction. The system doesn't consider more test data since it is not implemented with only four patients. The models are very simple and not able to handle complex data [20]. The sliding scale technique suggests insulin dosage with variations for different individuals without difficulty. It measures glucose based on the meal used by concerning a person [21].

More than one networking model is considered for prediction. If one model fails in prediction, the other model has the strength to predict the details. The ensemble methods are considered for prediction. None of the researchers consider illness and infection as a parameter [22].

The reinforcement learning (RL) framework is efficient in deciding within the timeline to observe the state of change of glucose controller. There is no need for training data for this model; the agent learns it through defined policies. RL algorithm is suitable for application with a time delay in response. The glucose generation and the insulin effects often have time delay in observation [23]. The polynomial-based multivariate data partitioning rule structure is used for the insulin dosage management process [34, 35]. HRDMR is a rule-based model, and it works based on divide and conquers structures [24, 25]. Utilizing the polynomial structures acquired through the techniques, forecasts are made for the patients whose drug dose values are unknown [26], but only fewer parameters are considered for type II drug analysis.

Despite the consequences performed with the UVA/PADOVA simulator, the method proposed by Ferran et al. [27] needs to be examined in the environment wherein elements aside from factors other than activities of physical and time of day are playing an effective role in glucose and insulin metabolism. This approach is needed to be examined with real-time clinical applications by considering humans with type I diabetes, because of the simulators not having the capacity to evaluate on a large scale.

Due to type II diabetes, the retina of the patients is damaged. It needs an early prediction and treatment. There are various segmentation techniques [41–48] used for analyzing the fundus images of the patients. The early diagnosis may save the life of the patients from vision loss.

Based on the literature, the cat boosting-based machine learning approach is proposed to analyze the type I diabetic patient's glucose level. There are multiple parameters such as physical activities, medical details, history of the patient, food, and other parameters that are considered for perfect prediction of insulin dosage. It is explained in the following sections.

9.3 Dataset

As shown in Table 9.2, for the prediction of insulin dose, the features of patients are considered. The dataset of insulin drug analysis consists of the following features: the age of each patient in years, the body mass index (BMI), the genetic details of the patient's parents diabetic status, insulin in the blood, C-peptide which is the by-product of insulin production in the pancreas, the blood sugar during fasting, urine

Table 9.2 Parameters of the dataset for insulin dosage prediction

Parameters	Range (min-max)
Age	40–83
Body mass index	21.60–40.50
Genetic	0–1
Insulin in blood (mcg/dl)	15.80–37.70
C-peptide	1–1
FBS (mg/dl)	87–448
Urine (mg/dl)	12–180
Creatinine (mg/dl)	0.60–4.20
Cholesterol (mg/dl)	87–480
Triglyceride (mg/dl)	106–639
HDL cholesterol (mg/dl)	21.60–78
LDL cholesterol (mg/dl)	25–399.40
Uric acid (mg/dl)	3.70–9.90
Acetone in urine (mmol/l)	0–0
Glucose in urine (mmol/l)	0–4

in blood, creatinine which is the waste product of creatine phosphate released from muscle and metabolism and will be in the blood filter through urine, cholesterol (mg/dl), triglyceride which is the fat in the blood, high-density lipoprotein which is made of protein and fat that carry cholesterol through the bloodstream, low-density lipoprotein (LDL), uric acid, acetone in urine, and glucose in urine.

9.4   Methodology

Nowadays the diabetes treatment is high. The proposed model using machine learning techniques reduces the cost of the treatment with affordable insulin pumps. Through the proposed system, the patients can monitor and control the diabetes for the short term as well as for the long term. The combination of different methods improves the advantages.

As shown in Fig. 9.1, the blood glucose is measured using CGM in the pancreas. The insulin required for digesting food is measured from the pancreas [28]. The required quantity of insulin will be injected through the insulin pump to maintain the proper insulin level in the human body to prevent from all other diseases.

It is a loop-based technique where after insulin supply in the bloodstream, the glucose is measured again to know the formation of glucose produced through this insulin. It enables the system to update the readings at regular intervals for monitoring the insulin necessary for the particular patient.

As shown in Fig. 9.2, the proposed method uses continuous glucose monitoring (CGM) devices such as the Dexcom G4 Platinum sensor. It can record the glucose level every 5 min. The android mobile application is classifying the food items in the meals, and it calculates the total calories on consumption. The people have to record the day to day exercises and stress level to measure the glucose level in their

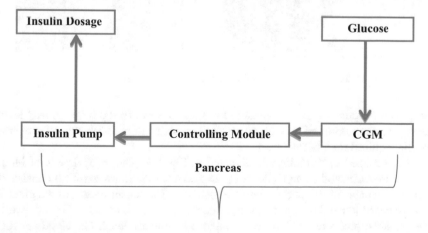

Fig. 9.1 Insulin management system in human body

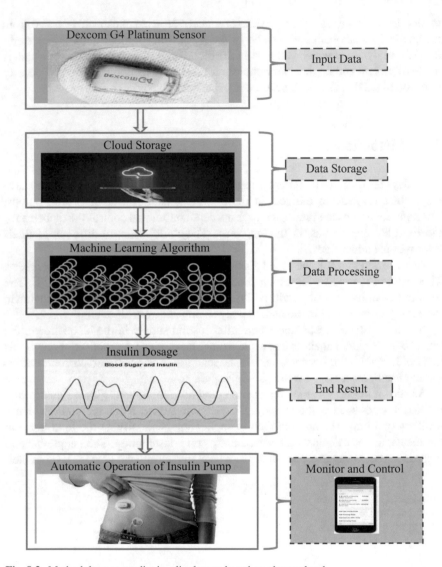

Fig. 9.2 Methodology to predict insulin dosage based on glucose level

body. The proposed system considers the glucose level in the blood, stress level, exercises, and consumed food based on all these parameters; it calculates the overall insulin required for a human being.

The proposed methodology is depicted in Fig. 9.2. The input details of blood glucose are measured using CGM of the Dexcom G4 Platinum sensor. It consists of a glucose sensor, Wi-Fi module, and a transmitter. The sensor measures the glucose level in the blood and transmits through the wireless medium to the cloud network. The private cloud is created to store the patient's glucose level. The cloud network contains the database to store the glucose readings every 5 min. The reading of the data is plotted as a graph in the mobile device. The android mobile application is connected to the central server to the cloud network. The sensor transmits data to the cloud through the server. The android mobile application displays the glucose value and the time using a graph. The graph clearly shows the value of glucose prediction at every moment.

The variations in the glucose level are observed easily through this graph. Based on the glucose level, the insulin required for the patients is calculated by considering other parameters also. This insulin is injected through the insulin pump whenever it is required for the individual. The machine learning approach provided an optimal solution in identifying the insulin required for specific individuals. The proposed paradigm improves the health of the patient and reduces the readmission and saves the economy of the country.

The blood glucose (BG) is measured through the calculation given as follows:

$$BGM = \frac{BG_{max} - BG_{min}}{BGC} \tag{9.1}$$

The objective of the proposed research is to provide the details of the bolus insulin dosage required for a diabetic patient. It is a rapid-acting insulin to increase blood sugar. It is calculated through the following mathematical calculation. The insulin dose is measured after carbohydrate is the ratio of total grams of carbohydrate disposed of in the meal to the grams of carbohydrate disposed of by one unit of insulin. The carbohydrate intake measurement (CHO) is depicted as follows:

$$CHO = \frac{CH_{max} - CH_{min}}{CFC} \tag{9.2}$$

CFC is the carbohydrate fitness coefficient. CH_{max} is the maximum carbohydrate intake, and CH_{min} is the minimal carbohydrate intake quantity.

In the case of eating 80 g of carbohydrate during a meal, then the CHO ratio is 1:10. To find the CHO insulin dose, it is the ratio of total grams of CHO in the meal (60 g) to the grams of CHO disposed of by one unit of insulin (10). It is 80/10. Therefore, eight units of rapid acting insulin are covered by the carbohydrate in the meal.

The insulin correction dosage is calculated as follows. The blood sugar correction dose is the difference between actual and target blood glucose to the correction factor. If there is one unit of blood glucose in the body which is 50 mg/dl, the high blood glucose correction factor is 50. The pre-meal blood glucose target is 120 mg/dl, and the actual glucose in the blood before the meal is 220 mg/dl. Hence the difference among the target is (200 − 120 = 100) and the insulin correction dosage is (100/50 = 2). Thus two units of rapid acting insulin are required to correct the blood sugar down to achieve the target of 120 mg/dl. Adding the carbohydrate coverage dose of eight units and a high glucose level correction dose of two units gives eight units of rapid-acting insulin needed for the body.

The total insulin requirement for the bodyweight is in pounds/4 or 0.55* weights in kg. Suppose the weight of the person is 80 kg, then daily insulin dose is $0.55 \times 80 = 44$ units. If the body is with higher resistance to insulin, then it needs low insulin and lower resistance to insulin which requires low insulin.

The blood glucose level of the patient is assessed using the heuristic method-based ranking system. This method finds the glucose level from the optimal blood glucose range.

$$BG_R = f_1 \left(BG - BG_{opt} \right) - f_2 \left(BG_h - BG \right) \tag{9.3}$$

In the above ranking function of BG, f_1 is the non-increasing function and f_2 is the non-decreasing function. BG_h is the hypoglycemia symptom among the low level BG patients. The normal BG must be 100 mg/dl for an adult.

Table 9.3 gives the details of the recommended insulin dosage for the patients under different test cases. *Case 1* is recommended for all general cases. For patients whose BG is not regulated by *Case 1*, *Case 2* is favoured. BG may be corrected by CABG, organ transplantation, or islet transplantation operations in which stable pancreatic beta cells are transplanted. Via *glucocorticoid* therapy, patients receiving high glucose can receive more than 80 units of insulin a day. *Case 3* is the patients who are not controlled by *Case 2*. *Case 4* is the patients not controlled by *Case 3*.

Table 9.3 Insulin infusions for intensive care to rapidly change the glucose level

Case 1		Case 2		Case 3		Case 4	
BG	Units per hour	BG	Units per hour	BG	Units per hour	BG	Units per hour
<70	0	<70	0	<70	0	<70	0
70–109	0.2	70–109	0.5	70–109	1	70–109	1.5
110–119	0.5	110–119	1	110–119	2	110–119	3
120–149	1	120–149	1.5	120–149	3	120–149	5
150–179	1.5	150–179	2	150–179	4	150–179	7
180–209	2	180–209	3	180–209	5	180–209	9
210–239	2	210–239	4	210–239	6	210–239	12
240–269	3	240–269	5	240–269	8	240–269	16
270–299	3	270–299	6	270–299	10	270–299	20
300–329	4	300–329	7	300–329	12	300–329	24
330–359	4	330–359	8	330–359	14	>330	28
>360	6	>360	12	>360	16		

9.5   Experimental Analysis of Cat Boost Model

The statistical study of the estimation of insulin dosage via the decision-making process. The cat boost model is one of the machine learning techniques of the gradient boost algorithm. This model is used for boosting up the decision trees.

9.5.1   Procedures of Cat Boost Model

The labelled characteristics are the input of the machine learning algorithm to classify the target characteristics which is shown in Fig. 9.3.

The classifier uses the following steps for making decisions:

1. The fitness function of the model.
2. Fine-tune the hyperparameters of the model.
3. Make predictions of insulin dosage.
4. Interpretation of results.

9.5.2   Mathematical Analysis of Cat Boost Model

The algorithm improves the predictions of other weak machine learning models. The leaves L of the tree T have right and left branches. These splits depend on the node and the path of the leaves in the search tree.

The split function depends on the feature F. If the split condition is positively met, then the object starts; it travels from the left subtree L_L. Otherwise, it starts from right subtree L_R. w_1, w_2 are the weights of the objects at the left and right leaves of the tree, respectively.

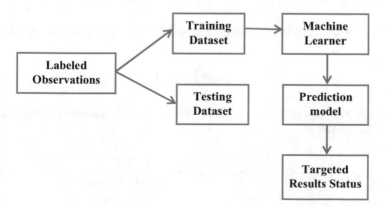

Fig. 9.3 The major steps of cat boost algorithms

The fitness function of this algorithm is defined as follows:

$$Core\_Feature_F = \sum_{T,L}(L_L - avg)''w_1 + (L_R - avg)''w_2 \tag{9.4}$$

$$avg = \frac{L_L''w_1 + L_R''w_2}{w_1 + w_2} \tag{9.5}$$

If the weight is not specified, then the weights depend on the number of objects in each leaf. The overall features of the leaf in the tree is defined as

$$Overall\_Feature_F = Core\_Feature_F + \sum_{i=1}^{N} avg\_Core\_Feature_F \tag{9.6}$$

However, $Core\_Feature_i$ is the individual feature of the leaves in the feature space. The $avg\_Core\_Feature_F$ is the average of all core leaf features in the ith combination. To achieve the best value of the feature, the ranking model-based calculations are used. This best feature metric value is obtained through the following computation:

$$Core\_Feature_i = \pm metric(C_i w) - metric(w) \tag{9.7}$$

The exact best feature metric value is computed as follows:

$$Core\_Feature_i = abs(metric(C_i w) - w_{best}) - abs(metric(w) - w_{best}) \tag{9.8}$$

$C_i w$ is the new leaf value set by the weighted average values of leaves with different paths.

$$C_i w = \frac{(n-1)w + C_F w}{n} \tag{9.9}$$

where the feature F is the combination of different leaves and it is represented as follows:

$$F = \{f_1, \cdots f_n\} \tag{9.10}$$

The random pool of all size of subset leaf in a tree is identified through the following computation:

$$L_{size} = min\left(Count, max\left(2e^5, \frac{2e^9}{Total\_Features}\right)\right) \tag{9.11}$$

The leaves in the tree with features F_1, F_2, and F_3 are interacting with other leaves by its connectivity. The direction of the interaction is calculated using eq. (12).

$$L\left(f_1, f_2\right)_{int\ eraciton} = \sum_T \left| \sum_{L_{Split(f_1)=Split(f_2)}} L_{value} - \sum_{L_{Split(f_1)\neq Split(f_2)}} L_{value} \right| \tag{9.12}$$

All the above mathematical expressions are used for finding the best decision to predict the insulin dosage based on the glucose level of the patients along with all other defined parameters.

9.6 Discussions

The proposed cat boot model is implemented over the dataset of the public as well as from real-time clinical values. The algorithm is prominent enough in predicting the insulin required for a patient daily.

For the experimental setup, STARR (STAnford Research Repository) dataset has been considered [49]. It consists of 42,700 patients details, in that there are 1.7% of people are having low blood glucose, 24.2% of people with high blood glucose, and 74.1% of people with normal blood glucose levels. Based on different filters applied over the database, 3461 patients are considered to train the algorithm. The remaining details are used for testing the algorithm. Apart from that, the clinical data of 352 patient's details have been collected from laboratories that are considered for validating the proposed techniques. The blood glucose level is updated based on consuming food which is the widely challenging part of the algorithm and that is the unique feature of the proposed method as well.

The statistical t-test evaluates the basic change in the results. The t-value is expected from the exhibition proportions of the proposed technique and competitive strategies. The result of the statistical t-test is given in Table 9.4. The acquired noteworthiness values for the test databases are 0.000 concerning the presentation measures. It has been reasoned that the proposed strategy is having a significant impact of the 0.05 level.

The parameters considered by the cat boosting algorithm finds the best value of match through tree search-based algorithm to take the average best decision. The optimal insulin dosage is selected through the efficient path navigation system of the algorithm. The rule-based policy improved the additional features considered in the system. The proposed technique achieved an average accuracy of 98.79%, sen-

Table 9.4 The statistical t-test on experimental data

Performance measures	Mean	Standard deviation	t-Value	Sig(2-tailed)
Sensitivity	99.83	5.25	48.09	0.000
Specificity	99.59	4.34	69.97	0.000
Accuracy	98.79	2.75	114.0	0.000

sitivity of 99.83%, and specificity of 99.59% in predicting the glucose to recommend the insulin dosage for type I diabetic patients.

9.7   Conclusions

The proposed diabetic disease analysis of type I diabetic patients considers the real-time data for disease prediction. Every time the patient's glucose level changes in their bloodstream is considered for analysis. The IoT devices enable the patients as well as physicians to diagnose the severity of the cases. The system is accurate enough to produce the required insulin quantity for the patients without erroneous value. The units of insulin suggested by the system consider the physical, medical, as well as food intake of the patients. The proposed approach overcomes the other approaches investigated by the researchers by considering the multiple parameters. The suggested insulin on a unit per day is evaluated by the experts against the laboratory evaluation. The required quantity of the insulin pumped through the insulin pump is also set to the manual mode of operations in some emergency situations. This helps the patients to get insulin dosage from a remote location without accessing the hospitals. The improved accuracy is the additional features of the cat boosting-based machine learning approach. Thus, the application of this research contributed to the healthcare systems to save the life of people in a long run as well as improve the economy of the country.

References

1. Insulin and Diabetes. https://www.diabetes.org.uk/guide-to-diabetes/managing-your-diabetes/treating-your-diabetes/insulin#:~:text=Insulin%20hel ps%20your%20body%20use,to%20take%20it%20as%20medication
2. Diabetic Ketoacidosis. https://www.diabetes.org.uk/guide-to-diabetes/compli cations/diabetic_ketoacidosis
3. Insulin, Medicines, Other Diabetes Treatments. https://www.niddk.nih.gov/health-information/diabetes/overview/insulin-medicines-treatments#medicines
4. Diabetes and Insulin. https://www.betterhealth.vic.gov.au/health/conditions andtreatments/diabetes-and-insulin
5. Cordera, R., & Adami, G. F. (2016). From bariatric to metabolic surgery: Looking for a "disease modifier" surgery for type 2 diabetes. *World Journal of Diabetes, 7*(2), 27–33.
6. Kirwan, J. P., Aminian, A., Kashyap, S. R., Burguera, B., Brethauer, S. A., & Schauer, P. R. (2016). Bariatric surgery in obese patients with type 1 diabetes. *Diabetes Care, 39*(6), 941–948.
7. IDF Diabetes Atlas. https://www.diabetesatlas.org/en/sections/worldwide-toll-of-diabetes.html
8. Pickup, J. C., Keen, H., White, M. C., Parsons, J. A., & Alberti, K. G. M. M. (1979). Long-term continuous subcutaneous insulin infusion in diabetics at home. *The Lancet, 314*(8148), 870–873.
9. Burleson, W., Clark, S. S., Ransford, B., & Fu, K. (2012). Design challenges for secure implantable medical devices. In *Proceedings of ACM 49th annual design automation conference*, 12–17.

10. Radcliffe, J. (2011). Hacking medical devices for fun and insulin: Breaking the human SCADA system. In *Proceedings of black hat conference presentation slides*.
11. Rathore, H., Al-Ali, A., Mohamed, A., Du, X., & Guizani, M. (2017). DLRT: Deep learning approach for reliable diabetic treatment. *GLOBECOM 2017-2017 IEEE global communications conference*.
12. Din, I. U., Guizani, M., Rodrigues, J. J. P. C., Hassan, S., & Korotaev, V. V. (2019). Machine learning in the Internet of Things: Designed techniques for smart cities. *Future Generation Computer Systems, 100*, 826–843.
13. Shifrin, M., & Siegelmann, H. (2020). Near-optimal insulin treatment for diabetes patients: A machine learning approach. *Artificial Intelligence in Medicine*.
14. Shokouhi, S., Sohrabi, M. R., & Mofavvaz, S. (2020). Comparison between UV/Vis spectrophotometry based on intelligent systems and HPLC methods for simultaneous determination of anti-diabetic drugs in binary mixture. *Optik-International Journal for Light and Electron Optics, 206*.
15. Murphree, D. H., Arabmakki, E., Ngufor, C., Storlie, C. B., & McCoy, R. G. (2018). Stacked classifiers for individualized prediction of glycemic control following initiation of metformin therapy in type 2 diabetes. *Computers in Biology and Medicine, 103*, 109–115.
16. Yazdjerdi, P., Meskin, N., Al-Naemi, M., Moustafa, A.-E. A., & Kovacs, L. (2019). Reinforcement learning-based control of tumor growth under anti-angiogenic therapy. *Computer Methods and Programs in Biomedicine, 173*, 15–26.
17. Li, N., Tuo, J., Wang, Y., & Wang, M. (2020). Prediction of blood glucose concentration for type 1 diabetes based on echo state networks embedded with incremental learning. *Neurocomputing, 378*, 248–259.
18. Franco, T., Aaronson, B., Williams, B., & Blackmore, C. (2019). Use of a real-time, algorithm-driven, publicly displayed, automated signal to improve insulin prescribing practices. *Diabetes Research and Clinical Practice, 156*.
19. Cui, S., Wang, D., Wang, Y., Yu, P.-W., & Jin, Y. (2018). An improved support vector machine-based diabetic readmission prediction. *Computer Methods and Programs in Biomedicine, 166*, 123–135.
20. Saiti, K., Macas, M., Lhotska, L., Stechova, K., & Pithova, P. (2020). Ensemble methods in combination with compartment models for blood glucose level prediction in type 1 diabetes mellitus. *Computer Methods and Programs in Biomedicine, 196*.
21. Introne, J., & Goggins, S. (2019). Advice reification, learning, and emergent collective intelligence in online health support communities. *Computers in Human Behavior, 99*, 205–218.
22. Woldaregay, A. Z., Arsand, E., Walderhaug, S., Albers, D., Mamykina, L., Botsis, T., & Hartvigsen, G. (2019). Data-driven modeling and prediction of blood glucose dynamics: Machine learning applications in type 1 diabetes. *Artificial Intelligence in Medicine, 98*, 109–134.
23. Tejedor, M., Woldaregay, A. Z., & Godtliebsen, F. (2020). Reinforcement learning application in diabetes blood glucose control: A systematic review. *Artificial Intelligence in Medicine, 104*.
24. Karahoca, A., & Alper Tunga, M. (2012). Dosage planning for type 2 diabetes mellitus patients using indexing HDMR. *Expert Systems with Applications, 39*, 7207–7215.
25. Fong, S., Mohammed, S., Fiaidhi, J., & Kwoh, C. K. (2013). Using causality modeling and Fuzzy Lattice Reasoning algorithm for predicting blood glucose. *Expert Systems with Applications, 40*, 7354–7366.
26. Torrent-Fontbona, F. (2018). Adaptive basal insulin recommender system based on Kalman filter for type 1 diabetes. *Expert Systems with Applications, 101*, 1–7.
27. Torrent-Fontbona, F., Massana, J., & Lopez, B. (2019). Case-base maintenance of a personalised and adaptive CBR bolus insulin recommender system for type 1 diabetes. *Expert Systems with Applications, 121*, 338–346.
28. Quiroz, G. (2019). The evolution of control algorithms in artificial pancreas: A historical perspective. *Annual Reviews in Control, 48*, 222–232.
29. Purushotham, S., Meng, C., Che, Z., & Liu, Y. (2018). Benchmarking deep learning models on large healthcare datasets. *Journal of Biomedical Informatics, 83*, 112–134.

30. Kintzlinger, M., & Nissim, N. (2019). Keep an eye on your personal belongings! The security of personal medical devices and their ecosystems. *Journal of Biomedical Informatics, 95.*
31. Tigga, N. P., & Garga, S. (2020). Prediction of type 2 diabetes using machine learning classification methods, International Conference on Computational Intelligence and Data Science (ICCIDS 2019). *Procedia Computer Science, 167*, 706–716.
32. Kavakiotis, I., Tsave, O., Salifoglou, A., Maglaveras, N., Vlahavas, I., & Chouvarda, I. (2017). Machine learning and data mining methods in diabetes research. *Computational and Structural Biotechnology Journal, 15*, 104–116.
33. Patrick Zeller, W., DeGraff, R., & Zeller, W. (2020). Customized treatment for Type 1 diabetes patients using novel software. *Journal of Clinical and Translational Endocrinology: Case Reports, 16.*
34. Dahad, N. *Blood-glucose monitor leverages machine learning for type 1 diabetes management.* https://www.embedded.com/blood-glucose-monitor-leverages-machine-learning-for-type-1-diabetes-management/
35. Contreras, I., & Vehi, J. (2018). Artificial intelligence for diabetes management and decision support: Literature review. *Journal Of Medical Internet Research, 20*(5).
36. Machine Learning for Managing Diabetes: 5 Current Use Cases. https://emerj.com/ai-sector-overviews/machine-learning-managing-diabetes-5-current-use-cases/
37. Sun, Q., Jankovic, M. V., & Mougiakakou, S. G. (2019). Reinforcement learning-based adaptive insulin advisor for individuals with type 1 diabetes patients under multiple daily injections therapy. *EMBC.*
38. Machine Learning Powers CDS Tool for Diabetes Management. https://healthitanalytics.com/news/machine-learning-powers-cds-tool-for-diabetes-management
39. Seo, W., Lee, Y.-B., Lee, S., Jin, S.-M., & Park, S.-M. (2019). A machine-learning approach to predict postprandial hypoglycemia. *BMC Medical Informatics and Decision Making.*
40. Daskalaki, E., Diem, P., & Mougiakakou, S. G. (2016). Model-free machine learning in biomedicine: Feasibility study in type 1 diabetes. *PLOS One.*
41. Jemima Jebaseeli, T., Anand Deva Durai, C., & Dinesh Peter, J. (2018). IOT based sustainable diabetic retinopathy diagnosis system. *Sustainable Computing: Informatics and Systems, 272.*
42. Jemima Jebaseeli, T., Anand Deva Durai, C., & Dinesh Peter, J. (2018). Retinal blood vessel segmentation from depigmented diabetic retinopathy images. *IETE Journal of Research.*
43. Jemima Jebaseeli, T., Anand Deva Durai, C., & Dinesh Peter, J. (2019). Extraction of retinal blood vessels on fundus images by Kirsch's template and Fuzzy C-means. *Journal of Medical Physics, 44*(21-26).
44. Jemima Jebaseeli, T., & Anand Deva Durai, C. (2019). Mechanism for diabetic retinal blood vessel profile measurement and analysis on fundus images. *Research Journal of Pharmacy and Technology, 12*(1), 1–6.
45. Francis, D. & Jemima Jebaseeli, T. (2016). Fundus image vessel segmentation using PCNN model. *Proceedings of 2016 online International Conference on Green Engineering and Technologies. IC-GET 2016.*
46. Jemima Jebaseeli, T., Sujitha Juliet, D., & Anand Deva Durai, C. (2016). Segmentation of retinal blood vessels using pulse coupled neural network to delineate diabetic retinopathy. *Communications in Computer and Information Science, 679*, 268–285.
47. Jemima Jebaseeli, T., Anand Deva Durai, C., & Dinesh Peter, J. (2018). Segmentation of type-II diabetic patient's retinal blood vessel to diagnose diabetic retinopathy. *Lecture Notes in Computational Vision and Biomechanics, 31*, 268–286.
48. Jemima Jebaseeli, T., Anand Deva Durai, C., & Dinesh Peter, J. Retinal blood vessel segmentation from diabetic retinopathy images using tandem PCNN model and deep learning based SVM. *Optik, 199*(2019).
49. STARR (STAnford Research Repository). https://med.stanford.edu/starr-tools.html

Chapter 10
Security Vulnerabilities and Intelligent Solutions for IoMT Systems

J. Jeyavel, T. Parameswaran, J. Mannar Mannan, and U. Hariharan

10.1 Introduction

The Internet of Medical Things (IoMT) is revolutionizing the medical field due to its various features like management of diseases and drugs, improving treatment methods and better patient care, and reduced cost. It can be visualized as an infrastructure of connected health monitoring devices along with software applications connected to healthcare through the Internet. In the USA, close to 70% of healthcare industry have already adopted IoMT systems [1]. However, due to the utilization of a large number and myriad types of devices for wireless transmission of medical data to the cloud, it introduces new challenges. Risks due to lack of security awareness among users, improper storage and transmission, device damage, denial of service (DoS), and stealing and manipulation of medical records are widely prevalent. IoMT can play an important role for the welfare of society during pandemic scenarios like COVID-19. However data security measures needs to be stringently employed so that the robustness and privacy of data is maintained [2].

J. Jeyavel (✉)
Department of Electronics and Telecommunication Engineering, Bharati Vidyapeeth College of Engineering, Navi Mumbai, Maharashtra, India

T. Parameswaran
Department of Computer Science and Engineering, Swarnandra College of Technology, Narsapur, Andhra Pradesh, India

J. M. Mannan
Department for Information Technology, MVJ College of Engineering and Technology, Bengaluru, Karnataka, India

U. Hariharan
Department of Information Technology, Galgotias College of Engineering and Technology, Noida, Uttar Pradesh, India

D. J. Hemanth et al. (eds.), *Internet of Medical Things*, Internet of Things,
https://doi.org/10.1007/978-3-030-63937-2_10

10.1.1   Existing Threat Analysis Frameworks and Shortcomings

Threat analysis or modeling is a technique of listing and grouping possible security challenges. The characteristics of existing frameworks such as STRIDE and LINDUNN are studied in brief and their analyzed shortcomings from the context of IoMT [3–5].

STRIDE
It is a widely cited framework, developed by Microsoft, majorly used in e-banking applications. STRIDE has been integrated with Software Development Life Cycle (SDLC) and uses data-flow diagrams which are used for mapping threats. STRIDE was a framework that was designing to withstand the following types of attacks:

- Spoofing
- Data tampering
- Data repudiation
- Information disclosure
- DoS
- Privilege misuse

Data-flow diagram (DFD) is used to study the complexity of the threat level, and based on it, threat ratings are assigned. The rating helps the security analysts to take appropriate countermeasures [4].

LINDUNN
It is a framework that deals with privacy issues in software industry. It is similar to STRIDE, as it uses data-flow diagram and gives a relationship table of possible threats with different components in DFD format.

The privacy threats are categorized as follows:

- Linkability
- Identifiability
- Non-repudiation
- Detectability
- Disclosure of information
- Unawareness
- Noncompliance

Other Approaches
There are various other approaches such as CORAS, quantitative threat modeling, abuser stories, STRIDE average tool, attack trees, fuzzy logic, and T-Map. Except for T-Map and CORAS, which use UML diagrams as a base artifact, all the other approaches use data-flow diagrams for modeling security threats [6–8].

The shortcomings of all the above threat models are that they do not address the issue of physical threat to the IoMT sensors, which are to be used at the user's end.

IoT/IoMT Device Models
An understanding of the IoT device communication framework is important for understanding their security requirements. IoMT systems differ from other computer networks due to their operating environments and various challenges since they are prone to physical attacks and can get damaged easily. Various models have been proposed at different times. Models such as three-layer model, five-layer model, and CISCO's seven-layer model were proposed for understanding IoT architecture. Among them, the three-layer model is not suitable since it does not take into account collaboration between devices. The five-layer model, which was proposed by the Academicia is also outdated since it does not provide scope for including mobile apps into the architecture [7–10].

The most suitable architecture framework that can be used to understand the rapidly developing IoMT scenario is the CISCO seven-layer model [10].

CISCO's Seven-Layer Model   This architecture allows for detailed understanding of the IoT architecture. Level one consists of basic sensors and other data capturing units such as camera. Level two denotes the connection between the nodes. Level three performs data processing. Level four deals with data aggregation and involves collecting data from various units and storage mechanisms. Level five involves abstraction of large amount of data. Level six involves data analysis of the abstracted data. Level seven provides the interface between the users and processes for collaboration. The transition from event-based computing model to a query-based system occurs during layer four and above. This model provides scope for better understanding the system and is widely followed in the industry and Academicia (Fig. 10.1).

10.2   IoMT Application Settings

IoMT can be visualized in three application settings [11–15].

Hospitals: IoMT is used for remote diagnostics from hospitals, predictive maintenance and performance upgrade, recycling, and waste management of various chemicals used in treatment of patients.

Home: IoMT system reduces the frequent intervention of a doctor for patients with chronic diseases and automatically alerting the health system, helping the elderly and disabled persons to remain in the comfort of their homes needing the 24-hr assistance of nursing personnel.

Body area network (BAN) sensors: These sensors help to monitor the user's health continuously and help to improve their health (Fig. 10.2).

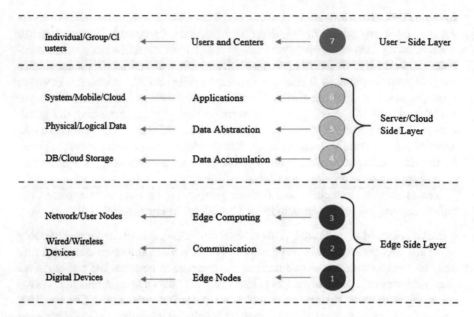

Fig. 10.1 CISCO seven-level model [8]

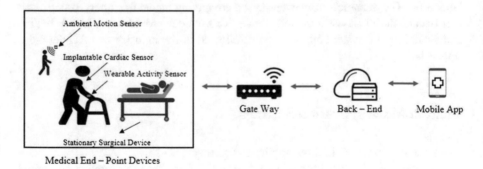

Medical End – Point Devices

Fig. 10.2 IoMT system [16]

10.3   IoT Architectures

10.3.1   Generic IoT Architecture

In a generic architecture, each device which is part of the communication ecosystem is given a unique IP address, and they can communicate with the common server using gateways. This type of architecture has provisions for different types of users, such as doctors and smart home user. Secure communication is facilitated by establishing a session key generated by cryptographic operations [15–18] (Fig. 10.3).

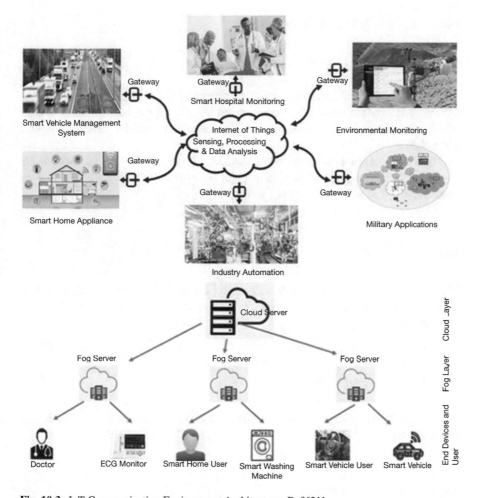

Fig. 10.3 IoT Communication Environment Architecture -Ref [21]

10.3.2 Fog-Based IoT Architecture

Due to the tremendous quantity of data generated by IoMT devices, the existing cloud platform was not able to address the security issues that come along with it. To overcome this challenge, CISCO introduced the concept of "fog computing" in 2012. In this, cloud servers process and manage the data close to IoT device such as a proxy server, thereby minimizing endpoint-to-endpoint delay of the network. Thus, it is also vulnerable to various attacks. To overcome this scenario, the nodes of the fog architecture interact with the adjacent nodes to keep watch for any attacker system [19–22].

10.4   Security Requirements

10.4.1   Authentication

There are two types of authentication – identity authentication and message authentication. Identity authentication validates the identity of the communication endpoints, whereas message authentication validates the message during message transmission.

The various entities such as smart devices, servers (cloud/fog servers), users (static/dynamic), and cloud service providers require authentication when they become part of IoMT environment [23–26].

10.4.2   Integrity

Message integrity needs to be maintained so as to prevent data tampering such as false insertion, unauthorized deletion, and modification during data storage and transmission.

10.4.3   Confidentiality

Confidentiality ensures protection to data from unauthorized users and guarantees privacy of data.

10.4.4   Non-repudiation

Non-repudiation refers to ensuring non-denial proof of communication to have occurred. This is made possible through the use of digital signature mechanism during online transaction. This mechanism registers the origin of the message, timeline of the message, as well as the destination of message. Thus the authenticity of the communication is guaranteed.

10.4.5   Authorization

This security mechanism refers to user privileges. A system administrator controls the access to the server and sets user privileges.

Table 10.1 DoS attacks possibility for routing protocol layer [26]

Layers	DoS attacks
Physical layer	Jamming, node tampering
MAC layer	Collision and unfairness
Network layer	Spoofing, replaying and wormhole, homing, hello floods
Transport layer	Flooding De-synchronization
Application layer	Overwhelming sensors, reprogramming attacks, route-based DoS

10.4.6 Newness

This assures that old message does not get transmitted and new or fresh message is transmitted at the latest instance.

10.4.7 Availability

Sometimes information may become inaccessible due to a special form of attack known as denial of service (DoS). This attack may prevent a registered use from accessing a server. Different types of DoS attacks are presented in Table 10.1.

10.5 Malware Attacks

Among the various security challenges faces by IoMT communication environment, malware botnets are the most malignant of the attacks, since they seriously underscore the privacy, robustness, and availability of the IoMT capabilities. Therefore, it is of paramount importance that IoT/IoMT environment is protected from this kind of attack. We have presented different types of malware attacks; their symptoms and a compilation of security protocols are provided. Most of the IoMT devices are fitted with wireless devices such as Bluetooth for endpoint-to-endpoint communication [27, 28].

10.5.1 Categories of Malware

Malicious software or in short malware is a malicious program that is transmitted over the network and is used to perform the following operations:

(1) Provide remote access to the attacker.
(2) Steal sensitive data from the affected system.

10.5.2 Symptoms of Malware

(1) Appearance of new and unrelated programs and icons on the device screen.
(2) The infected program may display irrational behavior like opening of multiple files, reconfiguring the system, and sending emails without the user's knowledge.
(3) A malware attack on a user end device of an IoMT system may tamper with user settings of myriad end devices such as oxygen level controller in a life support system [30–32].

10.5.3 Malware Types

Spyware This malware spies on the user activity and transmits information such as collected keystrokes, harvested data such as credit card details, etc.

Keylogger This code tracks the keystrokes of the user and transmits the user ID, password, etc. To overcome this, an intelligent user should not just rely on a strong password, but must use a combination of passwords and biometric access mechanisms such as face recognition software, fingerprint data, iris data, etc.

Trojan Horse This malware code resembles the behavior of a normal innocuous looking code and tricks the user into installing it and gains access into user system.

Virus This malware is able to copy itself and spread to other devices by connecting with to other programs, and if the user executes the program, it steals information or affects the executable files.

Worm It causes breakdown of the system by self-replicating itself and by consuming bandwidth and overloading the servers. They usually spread through email, and it does not need human intervention to be activated.

Adware It is a short form for "advertisement software." It opens up advertisements without the user's permission, and it tries to influence the user.

Ransomware This software is used to exhorting money from the device owner, by restricting the device owner's access to the system and making the system inaccessible (Fig. 10.4).

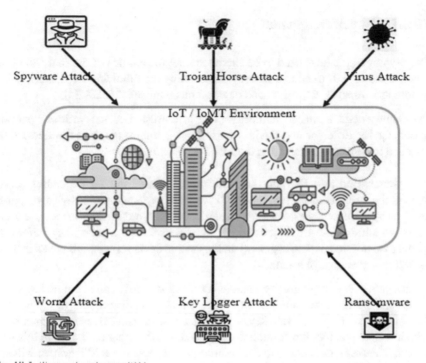

Fig. 10.4 Types of malwares [21]

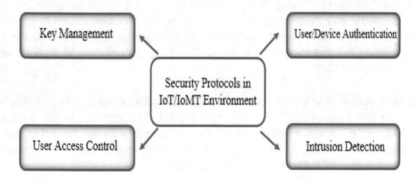

Fig. 10.5 Taxonomy of IoMT security management [27]

10.6   Taxonomy of Security Protocols

The IoMT security protocols can be categorized under the following heads (Fig. 10.5):

10.6.1   Key Management

The cryptographic keys need to be shared among the entities of the IoMT environment, and it is performed in a secure manner using the following steps, namely, key generation, transfer, activation, and deletion mechanisms [24, 29–34].

Pre-deployment a trusted authority (TA) is entrusted to set various operating parameter for different end devices performed in this stage, and the devices are installed in different parts within the IoMT environment.

Key Generation and Distribution During this phase, the TA depending on the need of the system creates two types of keys, namely, "symmetric key" and "public key" encryption schemes. In a "symmetric key" cryptographic scheme, the end users, which are part of the common process, exchange a secret key before the actual transmission of data begins. Usually, the end device credentials are exchanged as part of key exchange schemes.

In a public key cryptography scheme, a trusted authority called public key server is responsible for distribution of key. Both the communicating entities generate a pair of keys, namely, a private key, which is kept with themselves and a public key which is shared across the communicating media to the other endpoint. In this way, even if a hacker gains access to the common public key, the IoMT system will still be secure, since the hacker does not have access to the private key.

Key Establishment During this phase, the end device computes a session key based on the device parameters hidden within the visible message and is transmitted. This secret hidden message is verified by the end users, and based on mutual agreement, both the parties agree to transmit messages, since their credentials have been established [22].

Key Revocation Some time due to the physical threat or damage to a device, the existing private key stored in the device may need to be revoked, so that an unauthorized user does not capture it. Under these circumstances, the trusted authority (TA) revokes the existing private key and other confidential data and generates a new combination of public as well as private keys.

User Authentication During this process, the identities of the communicating parties are identified and verified. This can be done mutually between the end users and after the mutual authentication process; both the users exchange session key for future communication.

This process involves the following sub-processes [35–37]:

System Configuration The trusted authority, based on system parameters, registers the end user and, after successful registration, is deployed in the target area.

User Registration During this phase, the user registers him/her in a secure way and after registering sends his/her credentials using a secure channel to the trusted authority.

Login Phase During this phase, the administrator presents the credentials to a particular device, and the device performs the verification operation. If the credentials are matching, then the device sends a request message across the channel to the communication endpoint.

10.6.2   Authentication Phase

During this phase, the communication endpoint after receiving the request message from the source device initially verifies the authenticity of the message and generates a session key as part of authentication message and sends it back to the source device.

Password Update Phase The user authentication phase allows the user to change the password as per the need of the user, thereby increasing the security of the system.

Device Addition Phase This phase can be an optional phase which may be needed when the communicating device may get physically damaged or tampered with.

Access Control Access control refers to the ability of the user/device to access the resources of the system. System resources like medical data should not be protected from unauthorized access, and only privileged users should have the permission to access the system. It is also essential to safeguard accessibility since the adversary may install malicious code/device in the network. Thus, designing of secure access control mechanism are of utmost importance.

Access control mechanism can be implemented using the following schemes, namely, key establishment and node authentication.

Key Establishment
Secret key must be shared between newly installed node and its neighboring device to secure future communication.

Node Authentication In an IoT/IoMT environment, a new node should authenticate itself to its neighboring devices for future communication needs. Node authentication can be performed using two methods, namely:

Certificate-based access control mechanism: In this scheme, each deployed device must have a digital certificate that is preloaded in its memory. This will help to prove the node authenticity to its neighboring node.

Certificate-less access control mechanism: This method involves the implementation of a mechanism based on hash-chain to validate a newly joined endpoint's authenticity to the neighbors.

10.6.3   Intrusion Detection System (IDS)

An IDS system is used to detect the presence of malicious code/device within an IoMT system and to take remedial action based on the abnormality detected in the environment. An IDS continuously monitors the network traffic for signs of malicious activity. An IoMT system can be impacted through physical capture of IoMT nodes, extracting sensitive medical data from the device and replacing the captured device through a malignant node. These malignant nodes can be used to perform routing attacks such as blackhole, sinkhole, misdirection attacks, etc. This may result in the transmitted data packet to be delayed or destroyed during communication and can cause serious performance degradation of the network. Various network measurements such as throughput and packet delivery ratio tend to be affected adversely [38–42].

Network Intrusion Detection System An IDS mechanism identifies and reveals the nature of attacks carried out by an intruder node to the system administrator. Based on its working, an IDS system can be categorized into three different mechanisms:

(1) Anomaly-based detection
(2) Signature-based detection
(3) Specification-based detection

Anomaly-Based Detection This IDS scheme classifies network traffic flow as normal flow and abnormal flow based on statistical behavioral methods. Any change from normal behavior results in an alarm being raised by the system. The major advantage of this scheme is that it is useful for detecting new forms of malware activity in the IoMT environment and therefore useful for detecting new undocumented attack mechanisms. The challenge with maintaining this scheme is that it needs to be constantly updated form new legitimate behaviors of the system. This will prevent the occurrence of false-positive cases in the behavioral database.

Signature-Based Detection This scheme also called as misuse-based detection works on the basis of an existing database of known attack signatures. This detection scheme is the basis of most antivirus software and can detect anomalies effectively and accurately.

Specification-Based Detection This scheme combines the features of both anomaly-based IDS and signature-based scheme with the help of user-defined specifications to detect abnormal behavior at regular intervals to update the system. Manually defined specifications are used to validate the correctness of the detection process. This setup helps in reporting less false-positive cases as against anomaly-based mechanism.

10.7   Ontology-Based IoMT Security Model

Ontology refers to study and representation of data, entities, and their interactions. It can be used as a recommendation tool that models IoMT security system using various security threat scenarios as an input for decision-making. The ontology is supported with context-based rules for building a recommendation system that allows the stakeholders-end users and medical professionals to take informed decisions.

A recommendation tool that identifies various IoMT scenarios along with their countermeasures can be used. The tool highlights the challenges and risks involved in various solutions and guide them toward the most secure solution depending on their security considerations [43].

The problem is described in abstract mathematical and algorithmic forms with the following parameters:

Stakeholders It refers to the system administrator, patient, and medical professionals.

Solution It refers to device, service, and platform.

Component It refers to endpoint, gateway, mobile, and back-end requirement. It refers to security requirements such as data confidentiality, physical security, regulatory compliance, etc.

Issue It refers to a set of security issues that affect IoMT applications such unauthorized access, DoS, replay attacks, etc.

10.7.1   The Recommendation Tool

The proposed ontology-based tool works, on the basis of Python language that gets an IoMT user environment as input, identify the possible issues and recommend necessary relief actions The recommendation tool has the following characteristics [43–46]:

Ontology-Inspired The tool presents different use cases and models and integrates them into an IoMT-based real-world entity and relations.

Stakeholder-Centric The authors have identified three major stakeholders, namely, patients, healthcare professionals, and system administrators. Patients are the end users who utilize the wearable medical devices for monitoring their health conditions. Medical professionals include physicians, care takers, pharmacists, and technicians. System administrators are the professionals with technical expertise, such as IT professionals who are involved in building and maintaining the IoMT systems.

Description of Ontology Approach
The ontology-based approach describes the IoMT challenges and their security issues using four major components, namely [16]:

Concepts
Relationships
Instances
Axioms

Concepts They represent a set of main classes such as stakeholder, solution space, architecture, security measures, and attributes.

Instances They represent the things denoted by a concept. For example, "stakeholder" is an idea, and "patient" is a representation of the concept.

Relationships They represent the interaction between concepts. For example, the characteristics of the IoMT device type such as stationary, mobile, wearable, etc. are grouped under this component.

Axioms Axioms are used to indicate the relationship between concepts and their instances used as assertion to allocate for both concepts and instances.

An important aspect of the ontology-based approach is that it provides space for addressing security needs of specific scenarios. For example, an implantable medical sensor is less prone to stealing and may need a different security solution as compared to non-implantable devices. Thus ontology-based approaches have the potential to address various stakeholder needs.

10.8 Biometric-Based IoMT Security Systems

10.8.1 Unimodal Biometric System

Biometric identification systems can provide effective reliability compared to other identification systems since the password and keys cannot be stolen or falsified.

Most of the existing biometric systems rely on a single bio-feature such as iris, fingerprint, face, gait, voice, etc. However, a unimodal biometric system depending on a single bio-feature is prone to error due to various factors such as sensitivity to attacks, noise introduced in sensors, etc. These issues can be addressed by combining two or more biometric features and forming a fusion security system [47–49].

10.8.2 Multimodal Biometric System

Multimodal biometric system involves the usage of a combination of various biometric features and matching algorithms to overcome the drawback of unimodal systems. It reduces the possibility of error in this manner. The experimental results of the design framework have a higher recognition rate which ensures better performance than unimodal or single feature biometric systems.

The authors in [50] take into account a combination of biometric features as inputs. The advantage of fingerprint data is that they have high recognition rates. Similar is the case for face input data. However both of them may give imperfect result, if used alone, since they are susceptible to the quality of the image. In the same way, finger vein images are difficult to forge, but are not as precise as the other two. However, the authors point that by combining the attributes of the three types of data, i.e., by fusing the face, fingerprint, and finger vein data, they can compensate for the shortcoming of each, when taken individually. The fusion approach also avoids high cost and overcomes the need for multiple capture devices (Fig. 10.6).

The authors in [50] also have discussed various fusion levels from the matching score and decision viewpoints based on biometric features. At the individual level, feature vectors can be sourced from:

(1) Different sensors used for capturing a single biometric.
(2) Different entities based on a single biometric (e.g., left and right iris features).
(3) From multiple biometric traits.

Based on the different classifiers, matching-level or score-level fusion is proposed. At this level, combining of features can be done in two ways, namely, based on feature classifying problem or as a data combination problem. In the first method, based on the output of individual matches, feature vector is reconstructed. In the data combination approach, a single scalar score is generated from individual matching scores. The individual score is then normalized to a uniform field.

At the decision level, final decision is based on the combination of result matches obtained from different sets of matches. Various strategies such as majority voting, Bayesian technique, and Dempster-Shafer theory-based weighted voting are used at the decision level fusion. The authors have used PCA algorithm that uses reduced dimensionality to change high dimensional data to less dimensional data, to extract image features [41, 48–53].

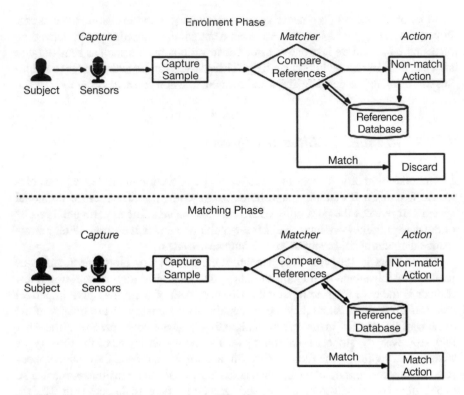

Fig. 10.6 Biometric authentication system [50]

10.9   Blockchain-Based Security Model

Blockchain is a decentralized, distributed storage technology which is behind the emergence of crypto-currencies. It stores information in the form of blocks and is a very secure mechanism. Blockchain technology can be used in medical servers since they need a heavy computational platform for generating blocks. Therefore this technology can be used in IoMT security mechanism for key management. It has been found to be resilient against DoS, replay, main-in-the-middle, impersonation, illegal key generation, etc. [54].

To improve the security of the blockchain mechanism for IoMT environment, the following data can be collected.

1. The end user's mobile device for authentication, since it contains important credentials
2. Password of the user
3. Biometric data from the user

The application of blockchain-based security system for IoT/IoMT applications is an active research area. In [55], a pioneering effort for storing and accessing public medical records in blockchain has been carried out. The blockchain-based approach provides security for backend cloud servers by preventing a range of attacks, namely, the replay attack, ephemeral secret leakage attack, impersonation attack, etc., and at the same time minimizes the computation cost of data transmission and processing.

10.10   Conclusion

The IoMT systems are likely to play a key role in providing affordable and easy healthcare for the future generations. IoMT systems can provide an intrusive and dynamic support for the healthcare system and is a boon to medical professionals. However, new security mechanisms need to be devised to keep the system robust and secure. In this chapter, we have presented the role of IoMT, its architecture, and various models that are being used. Malware poses a major challenge for IoMT systems, and they pose security and privacy challenges to the users. Various solutions such as using biometric systems for authentication of users, ontology-based security model for dealing with complex application scenarios, and blockchain-based security schemes for secure data access have been presented in this chapter. A summary of finding is presented in Table 10.2.

Table 10.2   Summary of various security models and their features [16, 50, 54]

Type of model	Stakeholders	Issue addressed	Security measure proposed
Ontology-based	Device, service users-medical professionals, patients, administrators	Malware Unauthorized access, Denial of service	Software security, access control, better authentication
Biometric-based	Patients, professionals, administrators	Authentication, accuracy	Efficient matching algorithm (Fischer vector), multimodal biometric schemes based on the face, fingerprint, and finger vein
Blockchain-based	Cloud servers, backend storage devices	Replay attack Man-in-the middle attack, ephemeral secret leakage attack, communication cost	Secure key management

References

1. HIPAA Journal. https://www.hipaajournal.com/87pc-healthcare-organizations-adoptinternet-of-things-technology-2019-8712
2. Dharavath, K., Talukdar, F. A, & Laskar R. H. (2013). *Study on biometric authentication systems, challenges and future trends.* In IEEE International conference on computational intelligence and computing research, pp. 1–7.
3. Hussain, S., Kamal, A., Ahmad, S., Rasool, G., & Iqbal, S. (2014). Threat modelling methodologies: A survey. *Sci. Int., 26*(4), 1607.
4. Möckel, C., & Abdallah, A. E. (2010) 2010 6th International conference on information assurance and security, IAS 2010, pp. 149–154.
5. Johnstone, M. N. (2010). *Threat modelling with stride and UML.* In Australian information security management conference (November), vol. 18.
6. Larry, G. (2007). *The security development lifecycle: Microsoft.*
7. Wuyts, K., Scandariato, R., & Joosen, W. (2016). *LINDDUN: A privacy threat analysis framework.*
8. Mohsen Nia, A., & Jha, N. K. (2016). A comprehensive study of security of Internet-of-Things. *IEEE Transactions on Emerging Topics in Computing, 5*(4), 1.
9. Aufner, P. (2019). The IoT security gap: A look down into the valley between threat models and their implementation. *International Journal of Information Security, 19*, 3–14.
10. Gubbi, J., Buyya, R., Marusic, S., & Palaniswami, M. (2013). Internet of Things (IoT): A vision, architectural elements, and future directions. *Future Generation Computer System, 29*(7), 1645.
11. Fleisch, E., Weinberger, M., & Wortmann, F. (2014). *Business models for the Internet of Things* (pp. 1–18). Zurich: Bosch IoT Lab.
12. Green, J. (2014). CTO data virtualization: IoT reference model white paper.
13. Challa, S., Wazid, M., Das, A., Kumar, N., Goutham Reddy, A., Yoon, E., & Yoo, K. (2017). Secure signature-based authenticated key establishment scheme for future IoT applications. *IEEE Access, 5*, 3028–3043.
14. Kumar, R., Zhang, X., Wang, W., Khan, R., Kumar, J., & Sharif, A. (2019). A multimodal malware detection technique for android IoT devices using various features. *IEEE Access, 7*, 64411–64430.
15. Wazid, M., Das, A., Odelu, V., Kumar, N., & Susilo, W. (2020). Secure remote user authenticated key establishment protocol for smart home environment. *IEEE Transactions on Dependable and Secure Computing, 17*, 391–406.
16. Alsubaei, F., Abuhussein, A., & Shiva, S. (2019). Ontology-based security recommendation for the Internet of medical things. *IEEE Access, 7*, 48948–48960.
17. Gatouillat, A., Badr, Y., Massot, B., & Sejdic, E. (2018). Internet of medical things: A review of recent contributions dealing with cyber-physical systems in medicine. *IEEE Internet of Things Journal, 5*, 3810–3822.
18. Kumar, P., Braeken, A., Gurtov, A., Iinatti, J., & Ha, P. (2017). Anonymous secure framework in connected smart home environments. *IEEE Transactions on Information Forensics and Security, 12*, 968–979.
19. Kumar, P., Gurtov, A., Iinatti, J., Ylianttila, M., & Sain, M. (2016). Lightweight and secure session-key establishment scheme in smart home environments. *IEEE Sensors Journal, 16*, 254–264.
20. Wazid, M., Das, A., Kumar, N., & Vasilakos, A. (2019). Design of secure key management and user authentication scheme for fog computing services. *Future Generation Computer Systems, 91*, 475–492.
21. Wazid, M., Das, A., Rodrigues, J., Shetty, S., & Park, Y. (2019). IoMT malware detection approaches: Analysis and research challenges. *IEEE Access, 7*, 182459–182476.

22. Wang, X., Wang, L., Li, Y., & Gai, K. (2018). Privacy-aware efficient fine-grained data access control in internet of medical things based fog computing. *IEEE Access, 6*, 47657–47665.
23. Yanambaka, V., Mohanty, S., Kougianos, E., & Puthal, D. (2019). PMsec: Physical Unclonable function-based robust and lightweight authentication in the internet of medical things. *IEEE Transactions on Consumer Electronics, 65*, 388–397.
24. Saleem, M., Mahmood, K., & Kumari, S. (2020). Comments on "AKM-IoV: Authenticated key management protocol in fog computing-based internet of vehicles deployment". *IEEE Internet of Things Journal, 7*, 4671–4675.
25. Shen, S., Huang, L., Zhou, H., Yu, S., Fan, E., & Cao, Q. (2018). Multistage signaling game-based optimal detection strategies for suppressing malware diffusion in fog-cloud-based IoT networks. *IEEE Internet of Things Journal, 5*, 1043–1054.
26. Das, A., Zeadally, S., & He, D. (2018). Taxonomy and analysis of security protocols for Internet of Things. *Future Generation Computer Systems, 89*, 110–125.
27. Sun, Y., Lo, F., & Lo, B. (2019). Security and privacy for the Internet of Medical Things enabled healthcare systems: A survey. *IEEE Access, 7*, 183339–183355.
28. Stallings, W. *Cryptography and network security*. Upper Saddle River: Prentice Hall Press.
29. Takase, H., Kobayashi, R., Kato, M., & Ohmura, R. (2019). A prototype implementation and evaluation of the malware detection mechanism for IoT devices using the processor information. *International Journal of Information Security, 19*, 71–81.
30. Rudd, E., Rozsa, A., Gunther, M., & Boult, T. (2017). A survey of stealth malware attacks, mitigation measures, and steps toward autonomous open world solutions. *IEEE Communications Surveys & Tutorials, 19*, 1145–1172.
31. Kumar, G. (2016). Denial of service attacks – An updated perspective. *Systems Science & Control Engineering, 4*, 285–294.
32. Kao, Y., Huang, K., Gu, H., & Yuan, S. (2013). uCloud: A user-centric key management scheme for cloud data protection. *IET Information Security, 7*, 144–154.
33. Li, J., Chen, X., Li, M., Li, J., Lee, P., & Lou, W. (2014). Secure deduplication with efficient and reliable convergent key management. *IEEE Transactions on Parallel and Distributed Systems, 25*, 1615–1625.
34. Tysowski, P., & Hasan, M. (2013). Hybrid attribute- and re-encryption-based key management for secure and scalable mobile applications in clouds. *IEEE Transactions on Cloud Computing, 1*, 172–186.
35. Jia Yu, Kui Ren, Cong Wang, & Varadharajan, V. (2015). Enabling cloud storage auditing with key-exposure resistance. *IEEE Transactions on Information Forensics and Security, 10*, 1167–1179.
36. Eschenauer, L., & Gligor, V. (2002). *A key-management scheme for distributed sensor networks*. In Proceedings of the 9th ACM conference on Computer and communications security – CCS '02.
37. Haowen Chan, & Perrig, A. (2003). Security and privacy in sensor networks. *Computer, 36*, 103–105.
38. Du, W., Deng, J., Han, Y., Varshney, P., Katz, J., & Khalili, A. (2005). A pairwise key predistribution scheme for wireless sensor networks. *ACM Transactions on Information and System Security (TISSEC), 8*, 228–258.
39. Blundo, C., De Santis, A., Herzberg, A., Kutten, S., Vaccaro, U., & Yung, M. (1998). Perfectly secure key distribution for dynamic conferences. *Information and Computation, 146*, 1–23.
40. Messerges, T., Dabbish, E., & Sloan, R. (2002). Examining smart-card security under the threat of power analysis attacks. *IEEE Transactions on Computers, 51*, 541–552.
41. Wazid, M., Das, A., Kumar, N., & Rodrigues, J. (2017). Secure three-factor user authentication scheme for renewable-energy-based smart grid environment. *IEEE Transactions on Industrial Informatics, 13*, 3144–3153.
42. Wang, D., Cheng, H., He, D., & Wang, P. (2018). On the challenges in designing identity-based privacy-preserving authentication schemes for mobile devices. *IEEE Systems Journal, 12*, 916–925.

43. Wazid, M., Das, A., Kumari, S., & Khan, M. (2016). Design of sinkhole node detection mechanism for hierarchical wireless sensor networks. *Security and Communication Networks, 9,* 4596–4614.
44. An Enhanced Privacy-Aware Authentication Scheme for Distributed Mobile Cloud Computing Services. (2017). KSII Transactions on Internet and Information Systems, *11*
45. Wenliang Du, Jing Deng, Han, Y. S., & Varshney, P. (2006). A Key predistribution scheme for sensor networks using deployment knowledge. *IEEE Transactions on Dependable and Secure Computing, 3*, 62–77.
46. Dolev, D., & Yao, A. (1983). On the security of public key protocols. *IEEE Transactions on Information Theory, 29*, 198–208.
47. Canetti, R. (2000). Security and composition of multiparty cryptographic protocols. *Journal of Cryptology, 13*, 143–202.
48. Canetti, R., & Herzog, J. (2010). Universally composable symbolic security analysis. *Journal of Cryptology, 24*, 83–147.
49. Roy, S., Chatterjee, S., Das, A., Chattopadhyay, S., Kumari, S., & Jo, M. (2018). Chaotic map-based anonymous user authentication scheme with user biometrics and fuzzy extractor for crowdsourcing Internet of Things. *IEEE Internet of Things Journal, 5*, 2884–2895.
50. Xin, Y., Kong, L., Liu, Z., Wang, C., Zhu, H., Gao, M., Zhao, C., & Xu, X. (2018). Multimodal feature-level fusion for biometrics identification system on IoMT platform. *IEEE Access, 6*, 21418–21426.
51. Pirbhulal, S., Wu, W., & Li, G. (2018). *A biometric security model for wearable healthcare.* In 2018 IEEE International Conference on Data Mining Workshops (ICDMW).
52. Challa, S., Das, A., Kumari, S., Odelu, V., Wu, F., & Li, X. (2016). Provably secure three-factor authentication and key agreement scheme for session initiation protocol. *Security and Communication Networks, 9*, 5412–5431.
53. Rajasegarar, S., Leckie, C., & Palaniswami, M. (2008). Anomaly detection in wireless sensor networks. *IEEE Wireless Communications, 15*, 34–40.
54. Garg, N., Wazid, M., Das, A. K., Singh, D. P., Rodrigues, J. J. P. C., & Park, Y. (2020). BAKMP-IoMT: Design of blockchain enabled authenticated key management protocol for internet of medical things deployment. *IEEE Access, 8*, 95956–95977.
55. Azaria, A., Ekblaw, A., Vieira, T., & Lippman, A. (2016). *MedRec: Using blockchain for medical data access and permission management.* In 2nd International conference on Open Big Data (OBD), pp. 25–30.

Chapter 11
Security Measures in Internet of Things (IoT) Systems Using Machine and Deep Learning Techniques

Teena Goud and Ajay Dureja

11.1 Introduction

The ongoing advancement in correspondence innovations, for example, the Internet of Things (IoT), has astoundingly risen above the conventional detecting of general conditions. IoT advancements can empower modernizations that upgrade the quality of life [1] and have the capacity to gather, evaluate, and comprehend the general conditions. This circumstance streamlines the new correspondence structures among things and people and subsequently empowers the acknowledgment of keen urban areas [2]. IoT is one of the quickest developing fields throughout the entire existence of figuring, with an expected 50 billion gadgets before the finish of 2021 [3, 4]. On the other hand, the crosscutting and enormous scope nature of IoT frameworks with different segments engaged with the organization of such frameworks have presented new security challenges.

IoT frameworks are intricate and contain integrative courses of action. Along these lines, testing is the key point in keeping up the security necessity in a large-scale assault of the IoT framework. Arrangements must incorporate comprehensive contemplations to fulfill the security prerequisite. Be that as it may, IoT gadgets generally work in an unattended domain. Thusly, a gatecrasher may truly get to these gadgets. IoT gadgets are associated ordinarily with remote systems where a gatecrasher may get to private data from a correspondence channel by listening in. IoT gadgets can't bolster complex security structures given their constrained calculation and force assets [5]. Due to the heavy use of IoT, there is need of interconnected undertaking by IoT organization. For instance, IoT frameworks ought to all the while think about vitality proficiency, security, large IoT information investigation techniques, and interoperability with programming applications [6] during the organization stage. One angle can't be disregarded when considering progresses in

T. Goud · A. Dureja (✉)
Department of Computer Science & Engineering, FET, PDM University, Bahadurgarh, India

© The Author(s), under exclusive license to Springer
Nature Switzerland AG 2021
D. J. Hemanth et al. (eds.), *Internet of Medical Things*, Internet of Things,
https://doi.org/10.1007/978-3-030-63937-2_11

another [7]. This coordination gives another open door for specialists from interdisciplinary fields to research ebb and flow difficulties in IoT frameworks from alternate points of view.

The IoT gadgets which give a huge and weak surface, there are need of additionally new security challenges. Due to which, important security issues arise. In addition, the IoT stage creates a huge volume of significant information. In the event that these information are not sent and investigated safely, at that point a basic protection break may happen.

IoT frameworks are open around the world, comprised for the most part of obliged assets and developed by lossy connections [8]. Thusly, pivotal alterations of existing security ideas for data and remote systems ought to be executed to give powerful IoT security strategies. Applying existing safeguard instruments, for example, encryption, validation, get to control, organize security and application security, is inadequate for Uber frameworks with numerous associated gadgets. For instance, "Mirai" is an extraordinary kind of botnets that has as of late caused hugescope DDoS assaults by misusing IoT devices [7, 9]. Existing security instruments ought to be upgraded to fit the IoT biological system [7]. In any case, the usage of security systems against a predefined security danger is immediately vanquished by new sorts of assaults made by aggressors to go around existing arrangements.

For instance, intensified DDoS assaults use caricature source logical addresses (IP address) for the assault area to be untraceable by safeguards. Subsequently, assaults that are more mind boggling and more ruinous than Mirai can be normal as a result of the weaknesses of IoT frameworks. In addition, understanding which strategies are appropriate for securing IoT frameworks is a test due to the broad assortment of IoT applications and situations [7]. In this way, creating viable IoT security strategies ought to be an examination need [7, 9].

As appeared in Fig. 11.1, having the ability to screen IoT gadgets can insightfully give an answer for new or zero-day assaults. AI and profound learning (ML/DL) are groundbreaking strategies for information investigation to find out about "typical" and "unusual" conduct. The information of each piece of the IoT framework can be gathered and examined to decide typical examples of communication, in this way distinguishing noxious conduct at right on time stages. Additionally, machine learning (ML)/deep learning (DL) strategies can be significant in foreseeing new assaults, which are frequently transformations of past assaults, since they can wisely anticipate future obscure assaults by gaining from existing models. Thusly, IoT frameworks must have a progress from simply encouraging secure correspondence among gadgets to security-based knowledge empowered by DL/ML strategies for viable and secure frameworks.

As the technique DL is the subfield of ML, this paper discussed these in two different areas to furnish the perusers with top to bottom audit, comprehensive examinations, and likely uses of both customary ML and DL strategies for IoT security. The primary contrasts between customary ML and DL strategies have been discussed in past writing [10, 11].

Fig. 11.1 Prospective view of IoT machine and deep learning in IoT

Likewise, in this paper ML alludes to customary ML techniques that require designed highlights, while DL strategies allude to ongoing advances in learning strategies that use a few non-straight preparing layers for discriminative or generative element reflection and change for design examination [12].

In this chapter, the ML/DL techniques are studies for IoT security. This security techniques can help specialists and designers as a manual for building up a viable and start to finish security arrangement dependent on insight. This review likewise means to feature the rundown of difficulties of utilizing ML/DL to make sure about IoT systems. Section 11.2 presents the IoT security properties and dangers and discusses the possible weaknesses and assault surfaces of IoT frameworks. In addition, we discussed another assault surface brought about by the IoT condition. In Sect. 11.3, we discussed the most encouraging ML and DL calculations, their focal points, drawbacks, and applications in the IoT security. Section 11.4 examines and exhaustively looks at the utilization of ML/DL strategies in making sure about each IoT layer. Section 11.5 presents the conclusions drawn from this study.

11.1.1 Related Work

A few scientists have directed studies on the IoT security to give a down to earth manual for existing security weaknesses of IoT frameworks and a guide for future works. Notwithstanding, a large portion of the current reviews on IoT security have not especially centered around the ML/DL applications for IoT security. For instance, the authors [13–19] inspected surviving examination and grouped the difficulties in the form of encryption, verification, get to control, organize security and application security in IoT frameworks; [20] stressed the IoT correspondence security subsequent to auditing issues and answers for the security of IoT correspondence frameworks; and [21] directed a review on interruption recognition for IoT frameworks. Weber [22] concentrated on legitimate issues and administrative ways to deal with IoT structures which fulfill the protection and security prerequisites. Roman, Zhou, and Lopez [23] discussed security and protection in the dispersed IoT setting.

In any case, rather than different overviews, our study presents a thorough audit of bleeding edge machine and late advances in profound taking in strategies from the point of view of IoT security. This overview recognizes and looks at the chances, preferences, and weaknesses of different ML/DL techniques for IoT security. We examine a few difficulties and future headings and present the recognized difficulties and future bearings based on surveying the possible ML/DL applications in the IoT security setting, in this way giving a valuable manual for scientists to change the IoT framework security from simply empowering a safe correspondence among IoT segments to start to finish IoT security-based clever methodologies.

11.1.1.1 State-of-the-Art Security Approaches of IoT

Assaults on and from IoT have gotten normal in the most recent years, for example, enormous scope denial-of-service (DoS) assaults on the Internet. This reality has incited a wide range of norms bodies and to give rules to designers and the Internet group everywhere to fabricate secure IoT and administrations.

(a) Correspondence Security: Correspondence in the IoT ought to be ensured by giving the security administrations examined in past segment. By utilizing normalized security components, we can give correspondence security at various layers.

(b) Information Security: Making sure about correspondence security is truly significant in IoT. But most of engineers making sure about information which is created from all IoT gadgets. The vast majority of gadgets in IoT are little and need more limitation because of restricted size to make sure about them from security dangers related with equipment.

11.2   Threats in IoT Security

Internet of Things incorporates the Internet with the world to give an astute connection between the physical world and on its environmental factors. For the most part, IoT gadgets work in various environmental factors to achieve various objectives. Nonetheless, their activity should meet a complete security necessity in digital and physical states [24, 25]. To fulfill the ideal security prerequisite, the arrangement ought to incorporate all-encompassing contemplations. Be that as it may, IoT gadgets generally work in an unattended domain. Thus, an interloper may genuinely get to these gadgets. IoT gadgets are ordinarily associated with remote systems where a gatecrasher may uncover private data from the correspondence channel by eavesdropping. Therefore, making sure about the IoT framework is a mind boggling and testing task. Figure 11.2 shows the potential assaults that can influence the primary security necessities.

11.2.1   IoT Threats

Violation of security can be arranged as digital and physical. Digital dangers can be additionally named inactive or dynamic. The accompanying subsection gives a concise conversation of these dangers.

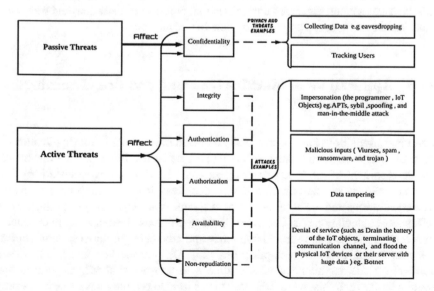

Fig. 11.2 Categorization of threats that assaults the security

11.2.1.1   Cyber Threats

Inactive Dangers   An aloof danger is performed distinctly by listening in through correspondence channels or the system. By listening in, an assailant can gather data from sensor devices and trace the sensor holders. Right now, gathering significant individual data, especially close to home well-being information, has gotten uncontrolled on the bootleg market [25]. The estimation of individual well-being data on the underground market is $50 contrasted and $1.50 for Visa data and $3 for a government-managed savings number [25]. In addition, an assailant can listen in on correspondence channels to follow the area of the Internet of Things gadget holder if it is correspondence.

Dynamic Threat   The assailant isn't just skillful in listening in on correspondence channels, yet additionally in adjusting IoT frameworks to change setups, control correspondence, refuse any assistance, etc. Assaults may incorporate a grouping of mediations, interruptions, and changes.

11.2.2   Assault Surfaces

In this segment, we discussed conceivable IoT framework assault surfaces and the potential dangers identified with each surface. IoT assault surfaces can be classified into physical gadget, organized administration, cloud administration, and web interface. Threat surfaces are presented by new IoT condition.

11.3   Applications of Machine Learning and Deep Learning in IoT Security

Learning calculations have been broadly received in some certifiable applications in view of their exceptional nature of taking care of issues. Such calculations handle the development of machine that progress naturally through experience [26]. The current headway of learning calculations has been driven by the improvement of new calculations and the accessibility of large information, notwithstanding the rise of low-cost calculations [26]. ML and DL have progressed extensively in the course of recent years, beginning from lab interest and advancing to down to earth apparatus with broad, noteworthy applications [26]. Despite the fact that DL field is a subfield of ML, in this chapter, ML alludes to conventional ML techniques that require built highlights, while DL strategies allude to ongoing advances in learning strategies that use a few non-direct preparing layers for discriminative or generative component reflection and change for design examination [12]. The reason for discussing ML and DL in two areas is to furnish the per user with inside and out survey of them two.

By and large, learning calculations plan to improve execution in achieving an errand with the assistance of preparing and gaining for a fact. For example, in learning interruption identification, the assignment is to arrange framework conduct as ordinary or strange. An improvement in execution can be accomplished by improving characterization precision, and the encounters from which the calculations learn are an assortment of typical framework conduct. Learning calculations are arranged into three principle categories: supervised, unsupervised, and reinforcement learning.

In this segment, we have discussed the most encouraging ML and DL calculations in security of IoT point of view. Right off the bat, we discussed the conventional ML calculations, their favorable circumstances, hindrances, and applications in IoT security. Furthermore, we discussed DL calculations, their favorable circumstances, detriments, and applications in IoT security.

11.3.1   Machine Learning Strategies for Security in IoT

11.3.1.1   Decision Trees (DTs)

DT-based strategies for the most part group by arranging tests as per their element esteems. Every vertex (hub) in a tree speaks to an element, and each branch signifies a worth that the node can be grouped in an example. Examples are grouped beginning at the root node and as for their component esteems. The component that ideally parts the preparing tests is regarded the starting point vertex of the tree [27]. Most DT-based methodologies comprise two principle forms: building (acceptance) and grouping (deduction) [20]. In the development (acceptance) process, a DT is developed ordinarily by at first having a tree with abandoned hubs and branches. Thusly, the element that best parts the preparation tests is considered the root node. This element is chosen utilizing various measures, for example, data gain. In the grouping (deduction) process, after the tree is built, the new examples with a set of highlights and obscure class are arranged by beginning with the root hubs of the built tree (e.g., the tree developed during the preparation procedure) and continuing on the way comparing to the scholarly estimations of the highlights at the inward points.

11.3.1.2   Support Vector Machines (SVMs)

SVMs are utilized for grouping by making a parting hyperplane in the information properties between at least two classes with the end goal that the separation between the hyperplane and the most adjoining test purposes of each class is expanded [21]. This technique is outstanding for their speculation ability and explicitly reasonable for datasets with countless element traits. Few example focuses in [22, 23]. Hypothetically, SVMs were set up from measurable learning [21]. They were at first made to sort straightly distinct classes into a two-dimensional plane involving straightly distinct information purposes of various classes (e.g., ordinary or irregular). SVMs should create an incredible hyperplane, which conveys greatest edge, by expanding the separation between the hyperplane and the most adjoining test purposes of each class. The upsides of this technique are their adaptability and their capacities to perform continuous interruption recognition and update the preparation designs progressively.

11.3.1.3 Bayesian Hypothesis-Based Calculations

Bayes' hypothesis clarifies the likelihood of an episode on the premise of past data identified with the episode [24]. For example, DoS assault location is related with arrange traffic data. In this way, contrasted and surveying system traffic without information on past system traffic, utilizing Bayes' hypothesis, can assess the likelihood of system traffic being assault by utilizing past traffic data. A typical ML calculation is dependent on Bayes' hypothesis. This hypothesis is called naive Bayes classifier.

11.3.1.4 k-Nearest Neighbor (KNN)

Euclidean separation as the separation metric is used by KNN classifiers [25]. It is a nonparametric technique. In Fig. 11.3, the red circles speak to vindictive practices, and the green circles speak to the typical practices of the framework. The recently obscure example (blue hover) should be named vindictive or typical conduct. The KNN classifier orders the new model on the premise of the votes of the chosen number of its closest neighbors; for example, KNN chooses the class of obscure examples by the greater part vote of its closest neighbors.

11.3.1.5 k-Means Bunching

k-Means bunching depends on an unaided ML application. This technique is used to find bunches in the information, and k alludes to the quantity of bunches to be produced by the calculation. This method is executed by iteratively assigning every information highlight from one of the k bunches as indicated by the given highlights. To generate an extreme output, the k-implies calculation is applied iteratively. Besides, after all the information tests are dispersed to a particular bunch, the group centroids are estimated by registering the mean of all examples. The calculation repeats these means until no example that can alter the groups exists [26, 27].

Fig. 11.3 Model of knn

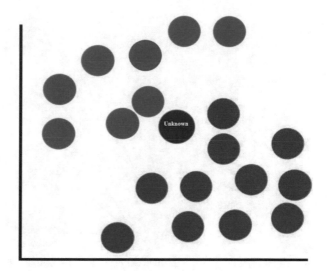

11.3.2 Deep Learning (DL) Techniques for Security in IoT

As of late, the utilizations of DL to IoT frameworks have become a basic examination point [28]. The most imperative favorable position of DL over customary ML is its boss execution in enormous datasets. A few IoT frameworks produce a lot of information; in this manner, DL techniques are appropriate for such frameworks. Additionally, DL can naturally separate complex portrayals from information [28]. DL strategies can empower the profound connecting of the IoT condition [40]. Profound connecting is a brought together convention that licenses IoT-based gadgets and their applications to associate with each other consequently without human intercession. For instance, the IoT gadgets in a brilliant home can consequently cooperate to shape a completely brilliant home.

DL techniques give a computational engineering that joins a few preparing levels (layers) to learn information portrayals with a few degrees of reflection. Figure 11.4 gives the knowledge of how the deep learning actually works.

11.4 Advantages of Deep Learning over Classical Data Mining

For traditional classical data mining methods, to successfully perform the data pre-processing as feature engineering is important. This process is based on selection of features. The main difference between deep learning approach and classical data mining method is that the former is able to perform most of the work necessary to

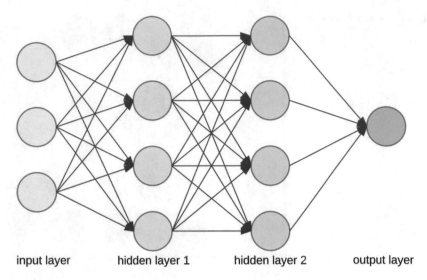

input layer   hidden layer 1   hidden layer 2   output layer

Fig. 11.4 Flow diagram of deep learning

form features using only consistently presented input data without the complex features which is selected manually.

In deep learning approach, there are cascading of layers. Each successive layer takes the output data from the previous layer as input data. These types of feature provide advantages over classical data mining methods when applied to variety of problems.

Computer vision, speech recognition, and natural language processing are the complex areas where the deep learning technologies are superior to the classical data mining methods and techniques. In deep learning approach, the accuracy level automatically increases, while the number of errors decreases.

In case of classical data mining approach, it presumes that humans develop functions, which is very time-consuming process. But deep learning is capable of generating new functions based on given dataset and can solve complex problems in an easier way.

11.5   Challenges, Issues, and Future Directions

In this area, we present a rundown of issues, challenges, and future headings for utilizing ML and DL strategies to moderate security shortcoming IoT frameworks, which are grouped dependent on information, learning systems, IoT situations, characteristic ML and DL challenges, chances to coordinated ML/DL with other innovation, computational multifaceted nature issues, and security versus different compromises necessities.

11.5.1 IoT Information-Related Issues

11.5.1.1 Availability of Security-Related Datasets

The main use of learning calculations is catching the designs from the accessible incomplete preparing dataset and afterward developing a model to order the new contributions on the premise of the scholarly examples. In this procedure, an inquiry to examine is the volume of preparing information required to prepare the learning calculations adequately for these calculations to be summed up for new contribution to the given space [26–28]. With regard to the utilization of ML and DL for IoT security, the significant test experienced by ML and DL, when all is said in done, and the administered ML and, what's more, DL strategies specifically, is the means by which to remove or produce a reasonable and top notch preparing dataset that contains different conceivable assault types. A great preparing dataset is a basic fixing to prepare the ML and DL calculations precisely. The preparation datasets should be extensive and differing. They ought to contain data that reflects about all of the procedures of certifiable assaults in light of the fact that these preparation datasets are the reason for getting model information. This condition can straightforwardly impact model exactness. Given that IoT frameworks create huge volumes of information, constant information streaming information quality upkeep stays a test.

An essential future exploration heading is the utilization of publicly supporting techniques for creating datasets identified with IoT dangers and assaults. Rich datasets that incorporate virtually all assault designs ought to be produced for preparing ML and DL. Besides, such datasets can be utilized to benchmark the exactness of recently proposed calculations against that of existing techniques for assault location. Despite the fact that creating communitarian IoT danger datasets, which can be persistently refreshed with new assaults, is critical, it is testing actually because of the wide decent variety of IoT gadgets. Moreover, a protection issue wins on the grounds that datasets may contain delicate or basic data that are not intended to be shared freely, explicitly for mechanical and clinical IoT gadgets.

11.5.1.2 Learning to Make Sure About IoT with Inferior Quality Information

The greater part of the proposed DL portrayals is by and large for excellent information. Be that as it may, IoT frameworks contain heterogeneous associated gadgets and enormous scope streaming, prompting the chance of high-clamor and ruined information to be assembled from such systems. Accordingly, figuring out how to secure IoT frameworks requires powerful DL models that can deal with and gain from inferior quality information, especially when acquiring top notch preparing information is for all intents and purposes infeasible.

11.5.1.3 Augmentation of Information of Security in IoT Enhanced Learning Calculation Execution

Instinctively, the more extravagant the information that ML and DL calculations have to gain from, the more precise they can be. In spite of the fact that getting a huge dataset is generally simple in specific spaces, for example, picture and characteristic language preparing, procuring a huge dataset for machine learning and Deep Learning is moderately troublesome in the area of information security in IoT frameworks. In this manner, discovering elective intends to acquire considerable measures of information in this space is attractive. Information growth is utilized to extend restricted information by producing new examples from existing ones. In the enlargement of IoT security information, the constrained measure of existing IoT security tests can be used to produce new tests.

11.5.2 Effective IoT Security Learning Strategies

11.5.2.1 Zero-Day Assaults on IoT

The primary bit of leeway of ML and DL strategies over conventional security strategies, for example, the danger signature-based strategy, is their ability to recognize zero-day assaults. Zero-day assaults, which are advancing dangers, were already unknown to discovery frameworks. These assaults have differing possibilities, such as transformative malware assaults that naturally reinvent themselves each time they are circled or communicated. Therefore, identifying these malware assaults by customary techniques is troublesome. The quantity of developing IoT security dangers [24], for example, zero-day assaults, is ceaselessly developing at a disturbing rate [24]. For instance, the Mirai botnet and also its deductions are turning into a disturbing danger to the security of IoT frameworks [7, 9]. The improvement of the ongoing determination of the Mirai botnet, Satori, demonstrates that other noxious IoT botnets are rising to abuse known and zero-day weaknesses.

11.5.2.2 Lifelong Learning for Learning IoT Dangers

One of fundamental attributes of the IoT condition is dynamism; a few new things join, and various articles leave the framework given the various and different IoT gadgets used to oversee various applications and situations. Given the IoT's nature, typical structures and examples of IoT frameworks may extensively change with time, and dangers and assaults focusing on the IoT framework may in like manner relentlessly fluctuate with time. Consequently, recognizing typical and unusual IoT framework conduct can't be consistently pre-characterized. Consequently, the visit refreshing of security models is required to deal with and comprehend IoT adjustments. In a genuine IoT condition, the transient learning of dangers and assaults focusing on IoT frameworks might be incapable for long haul security.

11.5.2.3 Transfer Learning

Move learning alludes to moving information from a space with adequate preparing information to an area with lacking preparing information. The primary reason for move learning is to decrease the time and exertion required for the new learning process. The primary worry in move learning manages the some portion of information that can be moved as information that is regular between the areas. Along these lines, moving such information is valuable. Then, moving information that is explicit for a specific area and doesn't hold any significance to different areas must be maintained a strategic distance form.

11.5.3   ML and DL for IoT Security in Related Interconnected and Intuitive Conditions

In this area, we present the open doors for utilizing ML and DL techniques to moderate interior security issues emerging from the structures of IoT frameworks, which are related, interconnected, and intelligent situations. As clarified already, with the quick increment in the number of IoT gadgets, the joint effort among gadgets is getting progressively self-sufficient; for example, they require diminished human contribution. IoT gadgets no longer just associate with each other like gadgets inside a system. Numerous current IoT gadgets are intended to accomplish the vision of a brilliant city, in which a large number of the gadgets are constrained by different gadgets or rely upon the operational state of different gadgets or the general condition. The upside of utilizing ML and DL in making sure about IoT gadgets in such a domain is that these techniques can be created to go outside essentially ability to grasp the operational conduct of explicit gadgets to understanding the operational conduct of whole frameworks and their gadgets.

11.6   Conclusion

The necessities for making sure about IoT gadgets have become complex on the grounds that few advancements, from physical gadgets and also remote transmission to versatile and cloud models, should be made sure about and joined with different advancements. The progression in ML and DL has taken into consideration the turn of events of different amazing systematic techniques that can be utilized to improve IoT security. In this review, different IoT security dangers and IoT assault surfaces are discussed. An exhaustive audit of the potential employments of ML and DL strategies in IoT security is given. These strategies are then looked at toward the finish of every subsection in terms of their points of interest, drawbacks, and applications in IoT security. A short time later, the employments of the ML and DL techniques for making sure about the principle IoT layers (e.g., observation, organization, and application layers) are assessed. At long last, a broad rundown of issues, difficulties, and future bearings identified with the utilization of ML and also DL in viably making sure about IoT frameworks are introduced and ordered by information; learning systems; ML and DL for IoT security in the reliant, interconnected, and intelligent conditions of IoT frameworks; assorted security exchange offs in IoT applications; and synergic mix of ML and DL with blockchain for IoT security.

References

1. Dastjerdi, A. V., & Buyya, R. (2016). Fog computing: Helping the Internet of Things realize its potential. *Computer, 49*(8), 112–116.
2. Yan, Z., Zhang, P., & Vasilakos, A. V. (2014). A survey on trust management for Internet of Things. *Journal of Network and Computer Applications, 42*, 120–134.
3. Evans, D. (2011). The Internet of Things: How the next evolution of the internet is changing everything. *CISCO White Paper, 1*(2011), 1–11.
4. Ray, S., Jin, Y., & Raychowdhury, A. (2016). The changing computing paradigm with Internet of Things: A tutorial introduction. *IEEE Design & Test, 33*(2), 76–96.
5. Abomhara, M. (2015). Cyber security and the Internet of things: Vulnerabilities, threats, intruders and attacks. *Journal of Cyber Security and Mobility, 4*(1), 65–88.
6. Serpanos, D. (2018). The cyber-physical systems revolution. *Computer, 51*(3), 70–73.
7. Bertino, E., & Islam, N. (2017). Botnets and Internet of Things security. *Computer, 50*(2), 76–79.
8. Kolias, C., Kambourakis, G., Stavrou, A., & Voas, J. (2017). DDoS in the IoT: Mirai and other botnets. *Computer, 50*(7), 80–84.
9. Xin, Y., et al. (2018). *Machine learning and deep learning methods for cybersecurity.* IEEE Access.
10. Chen, X.-W., & Lin, X. (2014). Big data deep learning: Challenges and perspectives. *IEEE access, 2*, 514–525.
11. LeCun, Y., Bengio, Y., & Hinton, G. (2015). Deep learning. *Nature, 521*(7553), 436.
12. Sicari, S., Rizzardi, A., Grieco, L. A., & Coen-Porisini, A. (2015). Security, privacy and trust in Internet of Things: The road ahead. *Computer Networks, 76*, 146–164.
13. Zarpelão, B. B., Miani, R. S., Kawakani, C. T., & de Alvarenga, S. C. (2017). A survey of intrusion detection in Internet of Things. *Journal of Network and Computer Applications, 84*, 25–37.
14. Roman, R., Zhou, J., & Lopez, J. (2013). On the features and challenges of security and privacy in distributed Internet of Things. *Computer Networks, 57*(10), 2266–2279.
15. Yaqoob, I., et al. (2017). The rise of ransomware and emerging security challenges in the Internet of Things. *Computer Networks, 129*, 444–458.
16. Wan, K., & Alagar, V. (2014). Context-aware security solutions for cyber-physical systems. *Mobile Networks and Applications, 19*(2), 212–226.
17. AlTawy, R., & Youssef, A. M. (2016). Security tradeoffs in cyber physical systems: A case study survey on implantable medical devices. *IEEE Access, 4*, 959–979.

18. Jordan, M. I., & Mitchell, T. M. (2015). Machine learning: Trends, perspectives, and prospects. *Science, 349*(6245), 255–260.
19. Faruki, P., et al. (2015). Android security: A survey of issues, malware penetration, and defenses. *IEEE Communications Surveys & Tutorials, 17*(2), 998–1022.
20. Das, S., Divakarla, J., & Sharma, P. (2015). *Detection and prevention of installation of malicious mobile applications*, ed: Google Patents.
21. Huang, J., Zhang, X., Tan, L., Wang, P., & Liang, B. (2014). Asdroid: Detecting stealthy behaviors in android applications by user interface and program behavior contradiction. In *Proceedings of the 36th international conference on software engineering* (pp. 1036–1046). ACM.
22. Kotsiantis, S. B. (2013). Decision trees: A recent overview. *Artificial Intelligence Review, 39*(4), 261–283.
23. Goeschel, K. (2016). Reducing false positives in intrusion detection systems using data-mining techniques utilizing support vector machines, decision trees, and naive Bayes for offline analysis. In *SoutheastCon, 2016* (pp. 1–6). IEEE.
24. Kim, G., Lee, S., & Kim, S. (2014). A novel hybrid intrusion detection method integrating anomaly detection with misuse detection. *Expert Systems with Applications, 41*(4), 1690–1700.
25. Alharbi, S., Rodriguez, P., Maharaja, R., Iyer, P., Subaschandrabose, N., & Ye, Z. (2017). Secure the internet of things with challenge response authentication in fog computing. In *Performance Computing and Communications Conference (IPCCC), 2017 IEEE 36th International* (pp. 1–2). IEEE.
26. Yerima, S. Y., Sezer, S., & Muttik, I. (2015). High accuracy android malware detection using ensemble learning. *IET Information Security, 9*(6), 313–320.
27. Bosman, H. H., Iacca, G., Tejada, A., Wörtche, H. J., & Liotta, A. (2015). Ensembles of incremental learners to detect anomalies in ad hoc sensor networks. *Ad Hoc Networks, 35*, 14–36.
28. Zhang, L., Zhang, L., & Du, B. (2016). Deep learning for remote sensing data: A technical tutorial on the state of the art. *IEEE Geoscience and Remote Sensing Magazine, 4*(2), 22–40.

Chapter 12
Mathematical Modeling of IOT-Based Health Monitoring System

G. Manoj, P. S. Divya, S. Raj Barath, and I. Justin Santhiyagu

12.1 Introduction

The mathematical modeling of Internet of Things (IoT) in the changing world has dramatically reduced the cost of the device design to the general public. The product of IoT [1] has created huge impact in the society in terms of accessibility to real-time applications such as household applications, transport, and day-to-day affairs. This paper emphasizes the technique for continuous monitoring of the patient in inaccessible areas. The data which has been continuously monitored from the patients has been transmitted [2]. In olden days, many different equipment's were used to monitor various parameters, but in this paper all the equipment's are integrated into one single module which performs the same task as the later does. IoT is the upgrowing technology in recent days. The Internet has become the vital part of one's life; in order to take this to the next step, the Internet has been connected to things. The use of smart technology is any device connected to an international

G. Manoj
Department of Electronics and Communication, Karunya Institute of Technology and Sciences, Coimbatore, Tamil Nadu, India
e-mail: manojzcbe@karunya.edu

P. S. Divya (✉)
Department of Mathematics, Karunya Institute of Technology and Sciences, Coimbatore, Tamil Nadu, India
e-mail: divya_deepam@karunya.edu

S. Raj Barath
Department of Electronics and Communication, Sai Vidhya Institute of Technology, Bangalore, Karnataka, India

I. Justin Santhiyagu
Department of Electronics and Communication, AJK College of Arts and Science, Coimbatore, Tamil Nadu, India

© The Author(s), under exclusive license to Springer
Nature Switzerland AG 2021
D. J. Hemanth et al. (eds.), *Internet of Medical Things*, Internet of Things,
https://doi.org/10.1007/978-3-030-63937-2_12

network called IoT. The data recorded by each sensor is uploaded to the server. All data uploaded to the server is protected by a logical ID and password. The dynamic innovation will have a transformative effect in each human's life and well-being checking; it will strikingly chop down the social insurance costs and a stride ahead in the exactness of ailment forecasts. In this paper, we present a thought of an assistance model in mechanical and monetary perspectives for the solace of patients and furthermore the open difficulties in actualizing IoT in certifiable clinical field. The framework ought to give the accompanying abilities: (a) capability of perusing the sign from the minimal effort PPG (photoplethysmographic) ear cut sensor, (b) providing the crude information over simple pin remotely over web attachment, (c) providing constant information handling registering of pulse and interbeat interval (IBI), (d) providing cloud administrations to help calculation code and UI for customers with different devices (PC, cell phone, keen TV, and so on), (e) multiple gadgets that can be associated with give effective following, and (f) providing standard GUI with intuitive key boundary illustrations.

12.2 Literature Survey

This is a well-demanded idea because of the development in areas like embedded, real-time communication, cloud computing, and data analytics. Gowrishankar S [2], a well-known professor, proposed a pulse and temperature sensing parameter using remote control. The wireless sensor senses the parameter data using web application and tracks the health of the patient regularly. All the sensed data is collected in the form of a database, and these data are used to notify the particular patient's diagnosis needed. Shreyaasha Chaudhury [3] has brought a revolution in the field of healthcare by continuously monitoring, aggregating, and effectively analyzing such information. The main aim is to accomplish a well-developed health analyzing system using IoT. In this system, it observes the various health frameworks and passes the information through a Wi-Fi module. Prema Sundaram [4] uses biological sensors to detect patient conditions and transmit data over the Internet. The sensors are interfaced to an Arduino board which is then connected to a LCD display for visualizing and reading the data generated. The data which has been read is stored and uploaded into the server, and finally it is converted into a JSON link for visualizing it on a smartphone.

According to Tarannum Khan [5], the system is primarily designed to monitor patients' vital signs such as heart rate, oxygen saturation, and plethysmogram. This device will be useful to physicians in constantly monitoring patients, providing information on their physical condition and providing them with limited mobility and greater comfort.

Extensive follow-up of patients is focused on several subgroups of patients: patients with persistent diseases, patients with transmission problems or other helpless patients, patients after medication, children, the elderly, and others. All of these patients are in a smarter state and need constant monitoring. The essence of good healthcare is to promote coexistence, as much fun as we can expect from all patients.

Most research follows a strategy that enables portability and mobility at home or in individual situations, which is valuable for the patient, and not to work in a doctor's office at significant prices. In the following, we will work on the entire structure to enable this idea to take advantage of several innovations.

The new application for remote monitoring of wellness enables elderly patients to participate in daily care without the help of a caregiver. In this way, these apps improve your practices such as sitting, standing, using the sink, watching TV, checking, and relaxing with minimal client workload. With or without a load-bearing sensor, these positions have minimal effect in practice. One of those models is a great clock-based sensor. The main elements of the remote monitoring structure are the information retrieval structure, the information collection structure, the clinical terminal, and the organization of communication. The information security framework consists of several sensors or modules that have built-in sensors with the ability to send remote information. As innovation advances, sensors are no longer just a clinical sensor. These can be cameras or cell phones. This is because continuous testing tests non-contact techniques in which the device does not touch the patient's body (McDuff et al. 2015).

The most basic type of these sensors utilized in with-contact techniques is remote sensor systems (WSN). These could be additionally arranged as remote body territory systems (WBAN), body zone systems (BAN), or individual region systems (PAN). Information handling framework incorporates a framework with information accepting and transmitting ability and a preparing unit/hardware. Terminal at the emergency clinic side can be either a PC (or a database) at the medical clinic, a committed gadget, or the smartphone of the specialist. Fundamental correspondence organize interfaces the information procurement framework to information preparing framework and further transmits the identified information and determinations to a human services proficient who is associated with the framework by means of the correspondence arrange.

In view of the unpredictability of the circumstance, the patient is either provoked to admit to an emergency clinic (by sending a rescue vehicle and so on.), do certain emergency treatment/alert advances, as well as take certain medications. The accessible far off well-being observing frameworks, their advancements, abilities, and activities shift to a huge degree. Istepanian [6] suggested a system which will diagnosis heart attack by calculating the heart rate by using IoT. In this method, they have used a pulse sensor, Arduino board, and a Wi-Fi module. The pulse sensor will start sensing the heart rate and will display the heartbeat of a person on LCD screen, and the data is transmitted using a Wi-Fi module. Also they have created an android application. This application [7] will track the heartbeat of a particular patient and monitor it correctly and give an alert message in chances of heart attack. Yedukondalu Udara [8] has developed a health monitoring system using Arduino microcontroller development board, Wi-Fi controller chipset, industrial grade temperature sensor, LCD display, and ECG shield. The ECG shields produce analog output, interfacing with Arduino Mega's analog pins. All sensor readings are read through the respective analog pins, and the variables are stored together with local display on the LCD. For interfacing with server, they used ThingSpeak IoT platform. The parameter readings are transmitted to the thing speak IoT platform using ESP8266 Wi-Fi interface.

According to Saraswati Saha [9], the system is primarily designed to monitor patients' vital signs such as heart rate, oxygen saturation, and plethysmogram. This device will be useful to physicians in constantly monitoring patients, providing information on their physical condition, and providing them with limited mobility and greater comfort.

Mohammad Salah Uddin [10] proposed an intelligent monitoring system with automatic sensor for monitoring purpose of patient's health condition. The biological behaviors of a patient are monitored using several sensors. For this insight and interconnection, IoT gadgets are furnished with installed sensors, actuators, processors, and handsets. IoT is anything but a solitary innovation; rather it is an agglomeration of different advancements that cooperate pair.

Sensors and actuators are devices that help detect physical conditions. The information collected by the sensor must be posted and processed at its discretion in order to draw useful conclusions from the sensor. Note that the term sensor is broadly characterized. Even a mobile phone or microwave oven can weigh the sensor if it contributes to the current state (internal state + state). A starter is a device used to influence the natural environment, such as a temperature regulator in a forced air system.

The capacity and preparing of information should be possible on the edge of the system itself or in a far off server. On the off chance that any preprocessing of information is conceivable, at that point it is ordinarily done at either the sensor or some other proximate gadget. The prepared information is then normally sent to a distant server. The capacity and handling abilities of an IoT object are likewise confined by the assets accessible, which are regularly exceptionally compelled because of impediments of size, vitality, power, and computational ability. Thus the fundamental examination challenge is to guarantee that we get the correct sort of information at the ideal degree of exactness. Alongside the difficulties of information assortment, and dealing with, there are difficulties in correspondence also. The correspondence between IoT gadgets is primarily remote since they are for the most part introduced at topographically scattered areas. The remote channels regularly have high paces of contortion and are untrustworthy. In this situation dependably, imparting information without such a large number of retransmissions is a significant issue, and consequently correspondence innovations are necessary to the investigation of IoT gadgets.

Presently, in the wake of handling the got information, some move should be made based on the inferred surmising's. The idea of activities can be different. We can straightforwardly change the physical world through actuators. Or on the other hand, we may accomplish something practically. For instance, we can send some data to other brilliant things.

The way toward affecting an adjustment in the physical world is regularly reliant on its state by then of time. This is called setting mindfulness. Each move is made keeping in thought the setting on the grounds that an application can carry on contrastingly in various settings, for instance, an individual dislike messages from his office to interfere with him when he is in the midst of a getaway.

Sensors, actuators, process, and communication servers are the main basis of the IoT system. In any case, there are many product perspectives to consider. First, you need intermediate software that you can use to process and manage the surfaces of these different parts. The interfaces of different modules require a lot of standardization.

The Internet of Things finds various applications in health services, wellness, education, entertainment, public activities, vitality protection, and condition monitoring, domestic mechanization, and transport structures. Section focuses on these areas of application. In each area of application, we can conclude that the development of IoT had a basic opportunity to reduce human effort and improve individual satisfaction.

The biological information is then transferred to the IoT cloud. Radosveta Sokulu [11] monitors steady-state blood pressure (KIT)-based blood pressure levels and healthcare feedback systems. Works with mobile application, with tight connection, inductive and magnetic connection. Once the KIT has been touched, the data is sent to a mobile app and to a secure website. Using this website, one can monitor patient's blood pressure level.

12.3   Modeling, Methodology, and Component

Mathematical modeling can be performed using three main methods; (1) threshold modeling, (2) clinical knowledge, and (3) preventive model. In the forecasting model, the type of perceived risk depends on the purpose of the risk forecasting; this could be for risk management or for identifying high-risk people, as Kaiser Permanente shows in Fig. 12.1. The stratification triangle has three levels: high, medium, and low. The data required for chance modification can be arranged in various manners. One potential categorization is the accompanying:

Segment attributes, for example, age, sexual orientation, inception, ethnic gathering
Clinical variables, for example, judgments and comorbidities
Financial attributes, for example, training, salary, or conjugal status
Well-being practices, for example, smoking, liquor utilization, and diet
Inclinations concerning personal satisfaction and desires on the medicinal services
    framework
Exploration Question: Combination of hazard factors

In the wake of recognizing the different hazard factors, the protected patients should be doled out to specific classes so as to assess them. Contingent upon the particular assessment and figuring model utilized, existing comorbidities will be consolidated in various manners.

Fig. 12.1 Risk
stratification triangle as
developed by Kaiser
Permanente

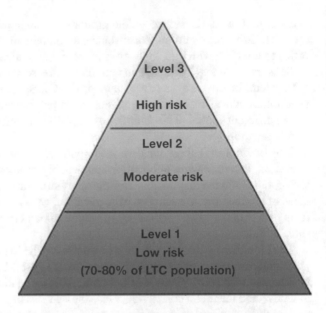

The second inquiry of intrigue was: "Which mix of hazard factors is utilized?"
This incorporates the subject of which strategies for chance change are as of now
utilized in universal social insurance frameworks to alter dangers and to anticipate
cost just as use.

Taking a gander at the consequences of the writing audit, it is by all accounts
sensible to make a separation by three levels:

Hazard modification techniques dependent on mortality dangers
Hazard modification strategies dependent on horribleness dangers, dependent on
    either pharmaceutical data or judgments
Hazard change with data on (self-revealed) well-being status

Another ordinarily utilized differentiation is made between the cell and the
relapse approach. On account of the cells approach, the patient is arranged by their
well-being components and morbidities to precisely one gathering. In light of the
data on the quantity of protected people in a single gathering and the normal treatment costs, the normal per capita costs for each gathering are determined. Along
these lines, the different gatherings of guaranteed are simpler to look at. Paradoxically,
in the relapse approach, a fundamental capitation expense, which depends, for
example, on the age and sexual orientation, is included with an additional factor
contingent upon the individual hazard variables of the person. Accordingly, a sup-
plier gets more additional charges for the treatment of a patient with many hazard
factors than for a patient with less dangers. The aggregate sum of repayment is
consequently the entirety of every single individual pay [3, 9, 10]. Hazard change
strategies are dependent on mortality dangers

The death rate is viewed as the most significant quality pointer. In any case, the
death pace of a medical clinic depends not just on the nature of the administrations

gave, yet is impacted by prior conditions, shifting degrees of seriousness of hidden ailments and malady free attributes, for example, age or sex. Along these lines, the clinical and monetary results of an emergency clinic additionally rely upon the hazard profiles of its patients. An emergency clinic rewarding high-hazard patients all the time has a higher mortality than a medical clinic with hardly any high-chance patients. A more regrettable result in such cases doesn't naturally infer a more unfortunate nature of care. Just a sensible thought of fluctuating dangers in the patient populace/case blend guarantees a reasonable correlation and may forestall preferences as far as referral designs and hesitance to work on high-chance patients. Also it might influence asset portion [12]. For this reason hazard balanced quality portrayals are utilized. So as to perceive and weight a potential normal impact of numerous hazard factors, calculated relapse models can be utilized.

Notable instances of mortality-based files are the Cardiac Surgery Reports, Charlson Comorbidity Index, or the Acute Physiology Age Chronic Health Evaluation (APACHE).

It is generally acknowledged that expenses must be considered in human services dynamic. The future conversation will concentrate on the nature of the medicines and the manners in which this quality can be surveyed and estimated. In medical procedure, usable or emergency clinic mortality is for the most part observed as the prime quality marker [10]. Be that as it may, the unadulterated thought of mortality is a fairly powerless proportion of value with regard to varieties for the situation blend of patients. In the event that employable mortality will be utilized as quality measure and hazard modification device, it must be changed in accordance with the fundamental case blend of a medical clinic [5].

12.3.1   Hazard Change Techniques Dependent on Grimness Dangers

By and large, a dreariness situated grouping is certainly not a clinical yet rather a cost-based order.

Not at all like supposed wordy-based models that are utilized for the figuring of per-case level rates for medical clinic releases (e.g., diagnosis-related groups, DRGs); order models for recipients are normally individual situated techniques. This incorporates the characterization depending on horribleness information for an entire year. In addition, asset usage is determined and gauge for a characterized timeframe (1 year). Consequently, evaluated costs are not arranged on a specific consideration period (as in DRGs) yet rather envelop the full scope of human services administrations and suppliers [13].

By and large, characterization models can be partitioned into models utilizing just data from one consideration segment, for example, outpatient medicine solutions, and those models utilizing similar pointers from various consideration parts, e.g., inpatient and outpatient analysis. These models are likewise called coordinated

218

G. Manoj et al.

models [14] and once in a while alluded to as "all-experience" models, as they include data from inpatient experience records to outpatient experiences [15].

12.3.2 Hazard Modification Strategies Dependent on Pharmaceutical Data

Pharmaceutical data can be viewed as a marker of constant illnesses [8]. So as to utilize this data, the dynamic specialists are relegated to infections regularly rewarded with the assistance of that medicate [16]. As the Anatomical Therapeutic Chemical Classification System with Defined Daily Doses (ATC/DDD) – framework by the World Health Organization [13] – gives a one of a kind characterization of restorative medications, the global likeness [17] is encouraged.

A bit of leeway of the utilization of pharmaceutical data is its finished accessibility. Information on solutions are regularly accessible in any event, when indicative information is missing [18, 19]. In addition, pharmaceutical data is not really manipulable and contrasted with other information [12]. Universally a few techniques are accessible. So as to contrast the various strategies and each other, a rundown of specific qualities has been characterized (creating foundation, targets, cell versus relapse approach, time of estimation, and so on.). Beyond what many would consider possible, the dissected strategies were surveyed as far as these. Table 12.1 gives a review on right now accessible strategies just as those utilized in past years.

Frequency range is between the minimum and maximum range.

Table 12.1 Heart rate values displayed on Ubidots

S.No.	Date/time	Values	Status
1	2020-4-20 16:19:55	89.94	Normal
2	2020-4-10 16:19:48	129.94	Abnormal
3	2020-3-12 16:19:42	89.83	Normal
4	2019-02-02 16:19:36	109.94	Abnormal
5	2020-01-20 16:19:30	90.01	Normal
6	2019-12-20 16:19:22	90.01	Normal
7	2019-11-20 16:19:17	89.85	Normal

Ubidots: In our task we are showing the data about quiet well-being-related boundaries like temperature and pulse in the site. Utilizing IP address, anyone can screen the patient's well-being status anyplace on the planet utilizing workstations, tablets, and PDAs. Ubidots website page is utilized for showing the data of the task

Numerical MODEL The methodology used in HMS was based on an improved dynamic system model. The global parameter is defined as μ, and any change in μ indicates that the condition of the cardiovascular system is improving. The parameter range is 0 <μ ≤ 1, so μ ≈ 1 refers to excellent cardiovascular health.

Taking these modifications into account, the following set of coupled ordinary differential equations expresses the improved model of heart rate kinetics:

$$HR\left(HR, HR\left(0\right), \lambda, v, t\right) = f_{min} f_{max} f_D \tag{12.1}$$

$$\dot{v} = I(t) \tag{12.2}$$

Improved methodology for mechanisms where the above Eq. 12.1.

A schematic model of the DVA, where m1 and m2 are the majority of the structure and DVA, individually, and k1 is the firmness of the essential structure and k2 is the solidness of the safeguard associated with the essential structure. We further expect that the structure has a straight gooey damping with damping coefficient c1, while c2 is the damping coefficient of the DVA. The removal of the structure is signified by x1, while the dislodging of the safeguard is meant by x2. We thought about two frameworks: one is with consonant power F(t)=F0cos(ωt) legitimately applied to the essential mass m1, and the other framework is agreeably energized by an uprooting x0(t)=F1cos(ωt) applied to the base of the essential framework, where we expect that F1=F0k1. Figure 12.1 Primary framework with gooey damped DVA (a) Force excitation framework and (b) Motion excitation framework

The conditions of movement of the framework with DVA for the symphonious power (condition 1) and base-excitation (condition 2) expect the structure

$$m1x''1 + c1x'1 + c2\left(x'1 - x'2\right) + k1x1 + k2\left(x1 - x2\right)$$
$$= F0\cos\left(\omega t\right) m2x''2 + c2\left(x'2 - x'1\right) + k2\left(x2 - x1\right) = 0 \tag{12.3}$$

$$m1x''1 + c1x'1 + c2\left(x'1 - x'2\right) + k1x1 + k2\left(x1 - x2\right)$$
$$= k1x0 + c1x'0m2x''2 + c2\left(x'2 - x'1\right) + k2\left(x2 - x1\right) = 0 \tag{12.4}$$

For our motivations, it is worth to revamp the conditions (12.4) and (12.5) in nondimensional structure, by consolidating the accompanying dimensionless boundaries

$$\tau = \omega 1t, u1 = x1xst, u2 = x2xst \tag{12.5}$$

where xst implies a static dislodging of the framework; u1 and u2 are the nondimensional relocations of the principle mass m1 and m2, separately; τ is the nondimensional time; and ω1 and ω2 are the characteristic frequencies of the essential framework and the safeguard, individually. With the replacement of the condition (3) into the conditions (1) and (2), the nondimensional type of the conditions of movement is acquired:

$$u''1 + 2\xi 1 u\dot{1} + 2\xi 2 \mu\varepsilon \left(u\dot{1} - u\dot{2} \right) + u1 + \mu\varepsilon 2 \left(u1 - u2 \right)$$
$$= A\cos\left(\Omega\tau\right) u''1 + 2\xi 2\varepsilon \left(u\dot{2} - u\dot{1} \right) + \varepsilon 2 \left(u2 - u1 \right) = 0 \tag{12.6}$$

$$u''1 + 2\xi 1 u\dot{1} + 2\xi 2 \mu\varepsilon \left(u\dot{1} - u\dot{2} \right) + u1 + \mu\varepsilon 2 \left(u1 - u2 \right) =$$
$$A\left(\cos\left(\Omega\tau\right) - 2\zeta 1\Omega\sin\left(\Omega\tau\right)\right) u''1 + 2\xi 2\varepsilon \left(u\dot{2} - u\dot{1} \right) \tag{12.7}$$
$$+ \varepsilon 2 \left(u2 - u1 \right) = 0$$

A=F0m1ω21xst,Ω=ωω1,μ=m2m1.

$$f_{min} \equiv \left[HR - HR_{min} \right]^{X}, \tag{12.8}$$

$$f_{max} \equiv \left[HR_{max} - HR \right]^{Y}, \tag{12.9}$$

X>0 and Y>0 are the parameter which controls model curve shape and the details of heart condition. The *fD* function that describes the attractor at D (λ,v,t) that can be defined as follows:

$$f_{D}\left(HR, HR(0), \lambda, v, t \right) = -d(\lambda)\left[HR - D(\lambda,v,t) \right], \tag{12.10}$$

12.3.3   Heartbeat Pulse Rate Sensor

Pulse is the speed of the heartbeat estimated by the quantity of constrictions (thumps) of the heart every moment (bpm). The pulse can change as indicated by the body's physical needs, including the need to retain oxygen and discharge carbon dioxide. It is normally equivalent or near the beat estimated at any fringe point. Exercises that can incite change incorporate physical exercise, rest, nervousness, stress, ailment, and ingestion of medications.

The American Heart Association expresses the ordinary resting grown-up human pulse which is 60–100 bpm.[1] Tachycardia is a quick pulse, characterized as over 100 bpm at rest [2]. Bradycardia is a moderate pulse, characterized as under 60 bpm very still. During rest a moderate heartbeat with rates around 40–50 bpm is normal and is viewed as typical. At the point when the heart isn't thumping in an ordinary example, this is alluded to as an arrhythmia. Variations from the norm of pulse some of the time show infection.

12.3.3.1   Materials and Technique

Initially, we will decide the contrast among heartbeat and heartbeat, and afterward we will talk about the technique and materials. For the distinction among heartbeat and heartbeat, as recently referenced, heartbeat is a unit for tallying the power of heart capacity and shows the pace of compression and extension of the heart, whose unit is pulsates every moment (bpm). The heartbeat can be estimated by an electro-cardiograph gadget. When blood goes through the heart and enters the veins, a weight wave is applied to the veins, which makes beats emerge in the veins. Truth be told, the beat is a withdrawal and extension in the veins that happens when the blood drops of the left ventricle. The beat can be contacted and estimated by setting fingers on the focuses where the vein goes through the bone (Fig. 12.2).

The primary concerns for estimating beat are outspread heartbeat and carotid heartbeat which the spiral heartbeat on the wrist and, what's more, the carotid heart-beat on the neck [10]. There is likewise a heartbeat meter gadget that can quantify the beat with infrared light to the ear cartilage. Under typical conditions, heartbeat is like heartbeat, since beat rate is identified with heartbeat. Figure 12.3 exhibits the heartbeat, and Fig. 12.4 shows beat rate outline.

12.3.3.2   Optical Screens

The heartbeat screen is a brilliant wearable gadget that distinguishes heartbeat from the body. This keen instrument employments photoplethysmography (PPG) innova-tion and furthermore have two sensors. The primary sensor is for identifying light and another for deciding movement. Its capacity is that the light is illuminated by the skin with a driven and afterward the light reflected from the body hits the locator and changes in heartbeat and body development are estimated. These optical screens use accelerometers to recognize movement. This gadget can be utilized during action, exercise, and rest or at some other time. The issues with these optical screens are the nearness of clamor in the gotten signals since they have comparative fre-quencies. Likewise, in a few cases, estimating heartbeat isn't right, which relies upon the physical body and the force of the client's activity.

Fig. 12.2 Heartbeat pulse rate sensor

Fig.
12.3 Temperature sensor

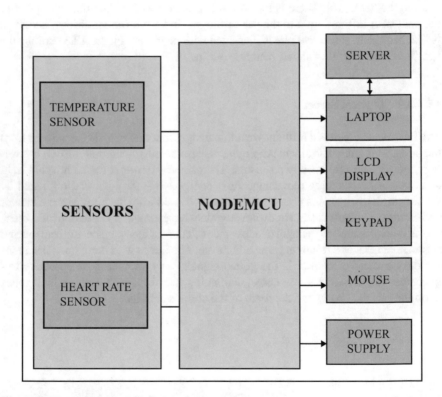

Fig. 12.4 Block diagram of the designed system

12.3.3.3 Segment Subtleties

The microcontroller is a component in the electronic circuits, which comprises a CPU, RAM, ROM, and info/yield ports installed in a little chip. Microcontroller is accessible in an assortment of gadgets including home apparatuses, cars, clinical gadgets, furthermore, more which is a controller inserted in these gadgets. The capacity of the microcontroller is that it gets the signs and afterward reacts as indicated by the sort of sign. Info and yield gadgets related with microcontrollers incorporate sensors, shows, switches as seen [14]. The microcontroller can get information with a sensor and at that point convert it to an electrical sign, and with this sign, it performs a particular capacity on the gadget. The PIC16F628A microcontroller which is shown in Fig. 12.5 was utilized in this examination. Heartbeat sensor is used to personally monitor the patient. It is used to measure and display the heart rate of the user. Heartbeat sensor [20] will return the digital output (from ADC) of the heart rate, once a finger is placed on it. The pulse will be detected, and the output can be sent to the microcontroller directly to measure the beats per minute (BPM).

Ubidots [21] is an IoT cloud platform which turns sensor data into useful information. This application platform supports interactive, real-time data visualization (widgets) and an IoT App Builder that allows developers to use the platform for private customization with their own HTML/JS code. Ubidots are used to enable visualization of data from the device. Through Ubidots we can produce continuous real-time graphs to find out the temperature and pulse rate [22] of the patients via wireless channel. The microcontroller [23] which we used for this project is NodeMCU which was programmed using C programming language in Arduino IDE with necessary libraries.

Fig. 12.5 Flowchart of the designed system

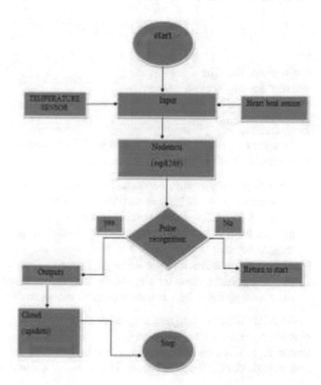

12.3.4 Temperature Sensor

The fundamental rule of working of the temperature sensors is the voltage over the diode terminals. In the event that the voltage expands, the temperature additionally rises, trailed by a voltage drop between the transistor terminals of base and producer in a diode. Other than this, Encardio-Rite has a vibrating wire temperature sensor that chips away at the standard of stress change because of temperature change.

The vibrating wire temperature meter is structured on the rule that divergent metals have an alternate direct coefficient of development with temperature variety. It principally comprises of an attractive, high elasticity extended wire, the two closures of which are fixed to any different metal in a way that any adjustment in temperature legitimately influences the pressure in the wire and, in this way, its normal recurrence of vibration.

The unique metal, on account of the Encardio-Rite temperature meter (Aluminum has a bigger coefficient of warm extension than steel), As the temperature signal is changed over into recurrence, a similar read-out unit which is utilized for other vibrating wire sensors can likewise be utilized for checking temperature too.

The adjustment in temperature is detected by the uniquely assembled Encardio-custom vibrating wire sensor and is changed over to an electrical sign which is transmitted as a recurrence to the read-out unit. The recurrence, which is relative to the temperature and thus to the strain 'σ' in the wire, can be resolved as follows:

$$f = 1/2 [\sigma g / \rho] / 2 l \, \text{Hz} \qquad (12.11)$$

where:
σ = strain of the wire
g = speeding up because of gravity
ρ = thickness of the wire
l = length of wire

Temperature sensors are accessible of different sorts, shapes, and sizes. The two fundamental kinds of temperature sensors are:

Contact-Type Temperature Sensors: There are a couple of temperature meters that measure the level of hotness or coolness in an article by being in direct contact with it. Such temperature sensors fall under the classification contact-type. They can be utilized to distinguish solids, fluids, or gases over a wide scope of temperatures.

Non-contact-Type Temperature Sensors: These kinds of temperature meters are not in direct contact of the article; rather, they measure the level of hotness or coolness through the radiation transmitted by the warmth source.

The DS18B20 is a temperature sensor as shown in Fig. 12.3 with the temperature reading from nine-bit to twelve-bits. The reading shows the temperature of a particular region. It uses one-wire bus protocol to communicate to the microprocessor and a single data line transmission. The sensor is directly powered by the data line;

hence additional power supply is not required. This DS18B20 is widely used in consumer product, industrial system, medical, agriculture, thermometers, and thermostatic controls.

12.4   Proposed System

The hardware implementation and working is proposed through the block diagram shown below. The findings are to elucidate that all the sensors operate properly without any loss of data. The pulse and temperature sensor extract the accurate readings and send them to the NodeMCU. This IoT device was able to learn the rhythm and measure the total temperature. The pace and temperature are constantly monitored and updated for the IoT section. The IoT section of this task is ThingSpeak.

The IoT arrangement created here is based on the Arduino ITA. The Arduino is one of the shortest and most famous prototypes. It is therefore recognized that the user has experienced the job of starting to use the Arduino and the Arduino LCD interface. The Arduino connects to the ESP8266 Wi-Fi modem, connects to a web switch, and accesses a cloud server. Arduino interfaces with a temperature sensor LM-35 that detects room temperature and a heart rate sensor that takes into account the heart rate. Intentional heart rate and body temperature are displayed on a character LCD connected to the Arduino and sent to the cloud phase, sending information to the Wi-Fi gateway. With this simple but successful tool, you can continuously monitor the health of a radically ill patient. It is usually used to monitor the strength of mature individuals who have heart or pulse problems.

Wellness information, such as pace and body temperature, will be updated at any time, and you will be linked to the ThingSpeak section. This information can also be used to preserve the patient's medical history. Freeboard.io is used as a panel to graphically display recorded information.

The Arduino sketch executed on top of the widget performs several task functions, such as checking the sensor information, converting it to wires, transitioning to the IoT phase, and displaying the estimated heart rate and temperature on the character's LCD. Sketches are created, collected, and pasted using the Arduino IDE. The IoT phase is used to create ThingSpeak, and Freeboard.io is used to create an IoT dashboard.

The Wi-Fi module (esp8266) which is also a part of the NodeMCU will relay the values to the server with no delay or data loss. The server will store all the data sent by the Wi-Fi module and display the same on the Ubidots web server and display the data in the 16×2 LCD display as well.

The pulse sensor amplified is a connection and playback pulse sensor for microcontrollers such as PIC, AVR, and Arduino. It is commonly used to easily integrate live heart rate information into your business. It basically connects to a basic optical pulse sensor that includes repair and flame-retardant hardware, making reliable heart rate measurements quick and easy.

Basically, cut the cartilage or fingertips of the ears and flexibly attach it to a 3.3 V or 5 V Arduino or battery. The time sensor module has three terminals: VCC, Ground, and Out. The shock on the impact sensor module is associated with a simple A0 needle on the Arduino. The VCC is connected to the Arduino 5 V DC output, and the ground is connected to a common display.

ESP8266 Wi-Fi Modem – The ESP8266 Wi-Fi module is used to pair your Arduino tablet with a Wi-Fi cloud access switch. This is an independent SOC with a coordinated TCP/IP convention set that allows access to Wi-Fi organizations. ESP8266 is ready to speed up your application or remove the entire capacity of the Wi-Fi organizer from another application processor.

Each ESP8266 module is pre-modified using the AT Order Kit software, which has two formats, ESP-01 and ESP-12. The ESP-12 has 16 pins, and these pins have access to the connector, while the ESP-01 has only 8 pins. The ESP-12 has a pin design.

The RESET and VCC module consoles are connected to the Arduino 3.3 V DC, while the ground pin is connected to a common display. The Tk and Rk contacts on the module are connected to pins 9 and 10 on the Arduino ITA.

16×2 LCD – Use a one-character LCD display to display the time speed and overall temperature. The 16×2 LCD screen is connected to the Arduino board, and the information pins are connected to the pins 3-6 of the Arduino board. The RS and E pins of the LCD are individually assigned to Chips 13 and 12 on the Arduino board. The LCD RV connector is grounded. The VCC terminal of the LCD module is connected to a 5 V DC Arduino. To resize the LCD module, connect the variable resistor to the VEE contact and place the other two variable resistor terminals between the VCC and ground.

Connect your LCD screen to the Arduino tablet in the task [24] using standard open source libraries. The library actually fills out a form, and no progress or customization is required. Power supply – 5 V DC power supply is required for all circuit components. The circuit is initially controlled by a 12V battery. The power applied by the battery is directed to the 5 V DC control voltage IC 7805. Pin 1 of the voltage regulator IC is assigned to the anode of the battery, and pin 2 of the battery is assigned to ground. Voltage performance is obtained from pin 3 IC.

In addition to the 10KΩ traction resistor, the LED is similarly connected between the split version and the firing pin to give a visual indication of normal progress. The LCD signal, impact sensor, and temperature sensor LM-35 receive 5 V DC power

12.5   How the Circuit Functions

This is an IoT-based pulse display model. It is very well used as a watch or earplug. Depending on the design of your wearable gadget, you can reject LCD characters and switch the entire circuit to a small control panel or SOC.

As soon as the circuit is powered by the battery, the Arduino will begin measuring the impact speed of the impact sensor and the ambient temperature of the LM-35 temperature sensor. The rhythm sensor has an infrared LED and a phototransistor that help detect shock at the end of the cartilage in the fingers and ears. Regardless of the rhythm of the distinguishing point, the IR LED has strips. The phototransistor recognizes the infrared LED light, and the obstacles change as the shock changes. A typical adult mind ranges from 60 to 100 at any given time. To identify each course (BPM), first define the obstacle that begins in two milliseconds. Therefore, the frequency of rhythm recognition control by the Arduino is 500 Hz, and this test result is sufficient to identify the heartbeat.

Along these lines, the Arduino simply checks the drive sensor voltage every 2 ms. A simple speed sensor output is converted to a computer estimate using a mounted ADC channel. The Arduino has 10 ADC channels, so the digitized values are between 0 and 1024. The focus is on the stimulus in this 512 range. First, the primary rate applies. If the status sensor has simple performance that stands out more than the central 512, then the following rates are the same. From now on, if the simple performance of the sensor is lighter than the center, all subsequent hits will be the same. For example, 512 and 3/5 after the time between beats recorded in the last cycle. The speed is detected and the variables addressed to BPM are updated without error. Due to this factor, this incentive is always sent to the fair and is used to discuss the actual heart rate or heart rate per minute. The Arduino code also uses the ability to blur the LED with each rhythm.

The sensor also identifies the internal heat level. LM-35 is used to differentiate temperatures. The operating temperature range of LM-35 is from -55 °C to 150 °C. The decay voltage differs by 10 mV depending on each increase/event °C in temperature. That is, its scaling factor is 0.01 V/oC. The LM-35 IC does not require any external adjustment or detachment to reach ± 0.25 °C at room temperature and ± 0.75 °C from -55 °C to 150 °C. Under typical conditions, the temperature estimated by the sensor will not exceed or reduce the operating range of the sensor. Typically, at temperatures from -55 °C to 150 °C, the output voltage of the sensor increases by 10 mV per degree Celsius. The voltage yield of the sensor is given by the accompanying formulae

$$\text{Vout} = 10\,\text{mV}\,/^\circ\,\text{C} * \text{T} \qquad (12.12)$$

where, Vout = Voltage yield of the sensor
T = Temperature in degree Celsius

$$\text{In this way, T}\left(\text{in}^\circ\text{C}\right) = \text{Vout}\,/\,10\,\text{mV} \qquad (12.13)$$

On the off chance that VCC is thought to be 5 V, the simple perusing is identified with the detected voltage more than 10-piece extend by the accompanying formulae

$$\text{Vout} = \left(5\,/\,1024\right) * \text{Analog} - \text{Reading} \qquad (12.14)$$

In this way, the temperature in degree Celsius can be given by the accompanying formulae:

$$T\left(\text{in}^\circ C\right) = \text{Vout}\left(\text{in V}\right)*100 \tag{12.15}$$

$$T\left(\text{in}^\circ C\right) = \left(5/1024\right)*\text{Analog} - \text{Reading}*100$$

In this way the temperature can be reasonably estimated by detecting a simple voltage at the sensor. The analogRead () task is used to detect simple voltages on contact pins. The Arduino collects information from both sensors and translates the quality into a string. Heart rate is a textual representation of the time between intentional heartbeat and rhythm, pronounced graphically according to the LCD sign. The temperature is also displayed on the LCD module.

The ESP8266 Wi-Fi module connected to the Arduino will send similar information to the ThingSpeak server when it detects a Wi-Fi access point. An advanced panel of experts or information is required to view and monitor the information sent to the ThingSpeak server. The company uses a panel of computers called Freeboard. io to filter sensor data on the Internet. Freeboard.io uses JASON records to record information from ThingSpeak.

It offers three components to construct a dashboard:

1. Data Sources – The information sources get the information from outside sources. These outer sources can be information merchant administrations, JavaScript applications, or JSON documents getting content from a HTTP server. In this task, the information source is a JSON document that gets information from the ThingSpeak server.
2. Widgets – The Widgets help to show information in literary or graphical structure. There are numerous gadgets accessible in Freeboard.io like content, diagram, measure, and so on.
3. Panes – These are utilized to compose gadgets for this module, which will be connected to the patient. These two input data will be sent to NodeMCU for further process. If the pulse is recognized, then the data will be sent to the web cloud server (Ubidots). If the pulse rate is not recognized, no data will be sent further (Figs. 12.6 and 12.7).

The hardware connections are made, and the code is uploaded in the NodeMCU board. With the temperature sensor [21], heart rate sensor starts reading the heartbeat and the temperature and displays the values in the LCD display. The values of both sensors are displayed in the LCD for every second. These values can also be monitored as a real time value in the Ubidots website. [20] The Ubidots website and the NodeMCU board are connected through

Wi-Fi. When the real-time values of heartbeat and temperature sensors are read, it will be displayed in the Ubidots website. Thus this can be helpful in monitoring the patients lively from anywhere through the cloud.

Fig. 12.6 Interface of the sensors and LCD display to the NodeMCU

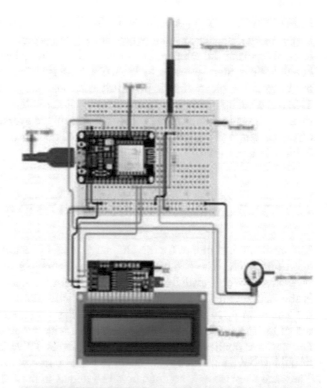

Fig. 12.7 Hardware implementation of the proposed system

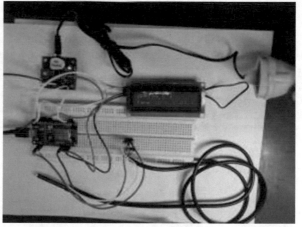

12.6 Experimental Results

12.6.1 Experimental Result on Temperature Heart Rate

The experimental result on healthcare system pulse rate of the with the Ubidots is done with health status of the system estimated by IOT system. The value of the heart rate of the IOT system is shown in Table 12.1. The heart rate display of the hardware implementation on the real-time system is done using Ubidots. The experimental result of the temperature value of the Ubidots health system based on the health of the system assessed by the IoT system is shown in Table 12.2. Piezo electric buzzer: The piezo electric bell is an electronic gadget used to produce a sound. This bell is set up by joining a piezo electric vibration plate in a plastic case (resonator). In this venture, this bell is utilized to caution the overseer during the basic state of the patient, and the sound got by this signal demonstrates that the patient wellbeing is in a hazard.

12.7 Mathematical Model Result on IOT HMS

The mathematical model for the predictive model is estimated for the health management system. The comparison of the mathematical model for clinical and the predictive model is evaluated. The percentage of the mathematical model of HMS is shown in Figure 12.8. The value of the predictive model is found to be better with the sample of up to 1000 peoples. The propelled arithmetic and numerical displaying of IoT is being driven by and utilized in a wide scope of scholarly, exploration, and business application zones. This utilization is delivering significant new functional involvement with a wide range of issue spaces in every one of these regions. There are likewise new computational strategies in these examination fields. As an empowering influence, this innovation is prompting a fast development in both logical and data applications that will, thusly, empower extra prerequisites for the propelled arithmetic and numerical demonstrating of Internet of Things to be recognized. These will affect the scholarly community, business, and training.

Table 12.2 Temperature values displayed on Ubidots

S.No.	Date/time	Values	Status
1	2019-11-20 16:19:55	71.13	Normal
2	2019-11-20 16:19:48	102.00	Abnormal
3	2019-11-20 16:19:42	72.40	Normal
4	2019-11-20 16:19:36	73.36	Normal
5	2019-11-20 16:19:30	93.50	Normal
6	2019-11-20 16:19:22	104.00	Abnormal
7	2019-11-20 16:19:17	72.00	Normal

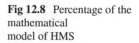

Fig 12.8 Percentage of the mathematical model of HMS

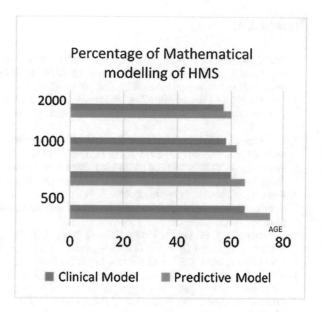

The real-time implementation of patient's healthcare is done using automated patient monitoring system. The proposed system monitoring of heartbeat as well as patient temperature is done automatically using cloud monitoring system. The heartbeat data monitor using Ubidots and the fluctuation in the heartbeat of a patient are intimated to the doctor using electronic mail. To monitor the heart rate, NodeMCU is used along with heart rate sensor using IoT platform.

The human temperature is measured with a DS18B20 temperature sensor which is used to monitor the patient's temperature and to apply the integrated system in the automation of the system in real time. Therefore, prototyping of patients' automated monitoring systems is supported and implemented. Automatic mathematical model provides the best performance of the health management system. With the wide utilization of web, this work is engaged to actualize the Internet innovation to set up a framework which would discuss through web for better well-being.

Internet of Things is required to manage the world in different fields however more advantage it would be in the field of social insurance. Subsequently present work is done to structure an IOT-based patient health monitoring framework utilizing a Raspberry pi3. In this work, the heart rate and body of the patient are observed utilizing Raspberry pi 3 and web server Ubidots. With the goal that the doctor address the issue using IOT HMS make concurrent appropriate move for the showed up issue. Subsequently constant patient health observing framework is planned.

References

1. Skraba, A., Koložvari, A., Kofjač, D., Stojanović, R., Stanovov, V., & Semenkin, E. Prototype of group heart rate monitoring with Nodemcu Esp8266. In *2017 6th Mediterranean conference on embedded computing* (Meco 2017, Bar, pp. 11–15).
2. Gowrishankar, S., Prachita, M. Y., & Prakash, A. (2017). IoT based heart attack detection, heart rate and temperature monitor. *International Journal of Computer Applications, 0975–8887*, 170–175.
3. Chaudhury, S., Paul, D., Mukherjee, R., & Haldar, S. *Internet of thing based healthcare monitoring system*. (978-1-5386-2215-5/17/$31.00 ©2017), IEEE.
4. Sundaram, P. (2013). Patient monitoring system using android technology. *International Journal of Computer Science and Mobile Computing, 2*(5), 191–201.
5. Khan, T., & Chattopadhyay, M. K. (2017). Smart health monitoring system. In *IEEE, international conference on information, communication, instrumentation and control (ICICIC-2017)*, paper id- 387.
6. Istepanian, R. S. H., Hu, S., Philip, N. Y., & Sungoor, A. (2011). The potential of internet of m-health things "m-IoT" for non-invasive glucose level sensing". In *33rd annual international Conference of the IEEE EMBS*, Boston, MA, USA (August 30–September 3, pp 5264–5266).
7. Kakria, P., Tripathi, N. K., & Kitipawang, P. (2015). A real-time health monitoring system for remote cardiac patients using smartphone and wearable sensors, Hindawi Publishing Corporation. *International Journal of Telemedicine and Applications, 2015*, 11–17. Article ID 37374.
8. Udara, Y., Udara, S., Harish, H., & Hadimani, H. (2018). Health monitoring system using Iot. *International Journal of Engineering and Manufacturing Science, 8*(1), 177–182. Issn 2249-3115.
9. Saha, S., & Majumdar, A. (2017, March). *Data centre temperature monitoring with Esp8266 based wireless sensor network and cloud based dashboard with real time alert system. Devices for integrated circuit* (Device, Kalyani, pp. 23–24).
10. Shaikh, S., Waghole, D., Kumbhar, P., Kotkar, V., & Jscoe, P. A. (2017, December). Patient monitoring system using IoT. In *2017 international conference on big data, IoT and data science (BID)* (pp. 20–22). Pune: Vishwakarma Institute of Technology.
11. Sokullu, R., Akka, M. A., & Cetin, H. E. (2010). Wireless patient monitoring system. In *Fourth international conference on sensor technologies and applications*.
12. Kumar, R., & Pallikonda Rajasekaran, M. (2016). Raspberry pi based patient health status observing method using internet of things. In *International conference on current research engineering science and technology (ICCREST-2016)*.
13. Navdeti, P., Parte, S., Talashilkar, P., Patil, J., & Khairna, V. (2019, March). Patient parameter monitoring system using raspberry pi. *International Journal of Engineering And Computer science, 5*(03), 16018–16021. ISSN:2319 – 7242.
14. Bhoomika, B. K., & Muralidhara, K N. (2017). Secured smart healthcare monitoring system based on IOT. *International Journal on Resent and Innovation Trends in Computing and Communication*.
15. Sivakanth, T., & Kolangiammal, S. (2016). Design of Iot based smart health monitoring and alert system. *International Journal of Control and Theory Applications, 9*(15), 7655–7661.
16. Giuseppe Mincolellim Gian Andrea Giacobone, Imbesi, S., & Marchi, M. (2020, June). Human centered design methodologies applied to complex research projects: First results of the PLEINAIR Project. In *International conference on applied human factors and ergonomics, AISC* (Vol. 1202).
17. Awaisi, K. S., Hussain, S., Ahmed, M., Khan, A. A., & Ahmed, G. (2020, June). Leveraging IoT and fog computing in healthcare systems. *IEEE Internet of Things Magazine, 3*(2), 52–56.
18. Gupta, S., Kashaudhan, S., Pandey, D. C., & Gaur, P. P. S. (2017, March). IOT based patient health monitoring system. *International Research Journal of Engineering and Technology (IRJET), 04*(03).

19. *Global challenges for humanity*. Available: http://www.millenniumproject.org/millennium/challenges.html
20. Lakmini, P. M., Ramzan, N., & Dahal, K. (2017). Remote patient monitoring: A comprehensive study. *Journal of Ambient Intelligence Humanized Computing*. https://doi.org/10.1007/s12652-017-0598.
21. Kumar, R., & Pallikonda Rajasekaran, M. *An Iot based patient monitoring system using raspberry Pi*.
22. Patel, N., Patel, P., & Patel, N. (2018). Heart attack detection and heart rate monitoring using Iot. *International Journal of Innovations & Advancement in Computer Science, 7*(4) Issn 2347 – 8616.
23. Mathan Kumar, K., & Venkatesan, R. S. (2014). A design approach to smart health monitoring using android mobile devices. In *IEEE international conference on advanced communication control and computing technologies* (ICACCCT, 2014, pp. 1740–1744).
24. Adivarekar, J. S., Chordia, A. D., Baviskar, H. H., Aher, P. V., & Gupta, S. (2013, March). Patient monitoring aystem using GSM technology. *International Journal of Multidisciplinary and Current Research, 1*(2).

Chapter 13
Machine Learning with IoMT: Opportunities and Research Challenges

M. Bharathi and A. Amsaveni

13.1 Introduction

Tremendous improvements in the technology over the last decade transformed the healthcare industry. Artificial intelligence (AI), machine learning (ML) and Internet of Things (IoT) are the key technologies which was responsible for this paradigm shift. All the healthcare devices are made smarter with multiple sensors embedded in it and are wirelessly connected to each other and also to the Internet. IoT enabling healthcare is an Internet of Medical Things (IoMT) [1]. There are five segments in IoMT [2]. 1.On-body segment: it includes wearable devices for health monitoring 2. In-home segment: this includes emergency response system, distance patient monitoring system mainly used to manage chronic diseases. 3. Community segment: this includes devices spread across the city, for example, mobility services and point of care device. 4. In-clinic segment: these are tools which are used for functional and therapeutic uses. 5. In-hospital segment: the devices present in the hospital for clinical and administrative use including product management, emergency management, resource management, etc.

One of the common benefits of IoMT with cloud storage is the possibility of accessing of the medical record by the healthcare professional and the patient at anytime from anywhere. This gives more flexibility to the healthcare professional and the patients as it removes the restrictions on the location. Exchanging of patients record with the expert clinician from remote places for better understanding of the possible illness becomes easy. Communication and controlling of the equipment in remote places are other advantage of IoMT [3].

M. Bharathi (✉) · A. Amsaveni
Department of Electronics and Communication Engineering,
Kumaraguru College of Technology, Coimbatore, India
e-mail: bharathi.m.ece@kct.ac.in

© The Author(s), under exclusive license to Springer
Nature Switzerland AG 2021
D. J. Hemanth et al. (eds.), *Internet of Medical Things*, Internet of Things,
https://doi.org/10.1007/978-3-030-63937-2_13

Diagnosis by extracting the insights of the medical record is one the real challenges in the healthcare sector. Diagnosis mainly depends on the knowledge and the experience of the physician. Cutting Edge Technologies can be leveraged for better diagnosis, thus ultimately delivering a better experience to patients. ML/DL algorithm is the current trend that helps the healthcare practitioners in assessing and analysing health data for improved diagnosis. A dynamic model with intelligence for data collection, analysis and prediction can be developed by using IoMT with ML/DL. Data generated by thousands of IoMT nodes within the healthcare system are collected using IoMT environment. Advanced analytics, communication and control within the connected devices benefit the stakeholders of the healthcare system. These benefits include more flexibility, seamless fusion of multiple technologies, continuous and personalized patient monitoring with wearable devices and big data analysis. Due to these advantages, ML is emerging as an integral part of IoMT.

13.2   Machine Learning for Data Analytics

The blending of IoMT with ML helps in better design of resource constraint IoMT devices. Use of AI and ML expands the applicability of IoMT. When ML algorithms work on IoMT, it is possible to achieve improved performance in terms of efficiency, accuracy and decision-making. Advanced monitoring of billions of sensing devices with improved communication between them strengthens the capability of IoMT and hence improves the quality of human life. This brings lot of growth in health sector. Figure 13.1 shows data analysis in IoMT using ML [1]. For ML algorithms to work efficiently, training data set should have a greater number of sample points. Huge volume of data collected using IoMT is analysed using machine intelligence to obtain deep insights from the data. This in-depth understanding of the data helps in making critical decisions for the real-life problems and starts the correct treatment on time. IoMT and ML should work together for better diagnosis and treatment. This area is highly researched in recent days due to the following reasons: IoMT accumulates huge amount of medical data which is highly valuable. This data needs to be analysed to extract useful information. Many data analytic methods and tools were used to infer knowledge from this data [5]. The hidden insights from the data can be effectively derived using ML methods with less human intervention. ML plays a vital role in constructing intelligent IoMT systems to analyse the data and to deliver timely services in the IoMT realm.

Data analysis can be classified in to descriptive analytics, predictive analytics and perspective analytics [4]. Descriptive analytics in IoMT environment gathers data from the IoMT devices and transfers to the cloud, and detailed insights into the past event are predicted using historical data by employing advanced ML techniques. For descriptive analysis, huge volume of data is needed which can be stored and processed in cloud. Predictive analysis uses historical data, and with ML techniques and advanced statistical model, the future trend will be predicted. Predictive

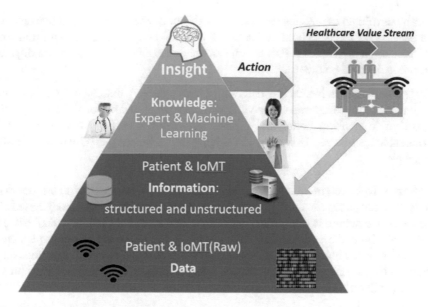

Fig. 13.1 Data analysis in IoMT using ML [1]

analysis is widely used in medical field for different applications. Prescriptive analytics examines the real scenarios and assist in decision-making [4]

Predictive analysis is most suitable for medical industry as it helps to get better insights of the medical record and hence helps in better and early diagnosis and timely treatment. At present many healthcare organizations move to ML-based predictive analysis methods.

13.2.1 Predictive Analysis

ML is used in predictive analysis to better understand the unknown parameters using huge volume of data. In medical industry, predictive analysis is very much needed as it helps to treat the patients with utmost by getting deeper insight about the record. This helps the health professionals to understand the problem at the early stage and take proactive measures to treat the patients. In predictive analysis, statistical tools and ML algorithms are used to analyse the historical data to realize the pattern. This knowledge is utilized to analyse the pattern in the new data. In summary, pattern in the new data is predicted using the existing data. Interconnection of medical devices helps to gather data which is needed for AI-based algorithms.

The success of predictive analysis in healthcare sector depends on identification of proper data set to be used for an application, obtaining quality data and applying suitable model for the analysis. For successful predictive analysis model, the sample size should be in millions. Smaller sample size will affect the accuracy of finding.

This huge volume of medical data set is possible only with IoMT. Predictive analytics benefits all the stakeholders including patients, hospitals, insurance providers and product manufacturers. Few benefits of predictive analysis in healthcare organizations is detailed below [6].

Early Diagnosis In healthcare, detecting and treating of a disease before it creates huge effect is very important. The volume of data generated by IoMT helps the ML algorithm to perform better and predict the illness accurately before it shows remarkable symptoms. This helps in deciding the treatment plan and saves the life of people.

Research Research in the area of predictive analysis is gaining its importance due to its numerous advantages. Any abnormality detection in the human body based on the data set available is made more accurate with the help of predictive analysis. In drug trials, limited set of people will be tested for any new drug before implementing for larger population. The effect of the drug on the patients will be effectively analysed by gathering large set of timely data. The outcome of this helps in improving the quality of drug and treatment plan.

Quality of Life Care Based on the availability of complete data from various IoMT devices and analytical tools, customized treatment can be planned for patients, and hence the quality of healthcare improves.

Cost Savings Cost saving to health providers is possible using optimization process by predicting admission rate, requirement of doctors, medicines, etc. Based on these predictions, the infrastructure and facilities can be planned. Predictive analysis can also be used for identifying patients likely to skip the appointment without notice. This understanding helps in planning the appointments properly to improve patient satisfaction.

Improving Operational Efficiency ML can predict the trend in the incoming patient numbers and seasonal sickness patterns. This helps hospitals to plan for additional facilities and also helps to maintain optimum patient to staff ratio.

Personal Medicine Planning treatment for each individual separately for certain types of illness is the trend today. Some medicines may work good for some people and not for others. Personalized treatment improves the satisfaction and the result. Analysing all the symptoms of each individual patient can be done effectively by predictive analysis. Predictive analysis can find the correlation between the symptoms and can bring out the hidden pattern by analysing huge volume of data and thus helps in planning personalized treatments.

Estimation of Risk Score Predictive analysis can be used to estimate the risk score of individuals using big data. At initial stage of disease progression, PA can be used to estimate the probability of developing chronic conditions.

Outbreak Prediction ML and AI are the effective tools used to understand the spread of contagious disease in the society. This can also identify the scale of spread, spread pattern in different regions, etc. This helps in planning different measures to curb the spread. The prediction of COVID-19 spread was also estimated by AI-based algorithms.

Controlling Patient Deterioration Possible deterioration of patients based on the vital parameters can be understood before they actually manifest. This helps in treatment planning.

Supply Chain Management Predictive analysis also helps in decision-making for negotiating the price, optimizing the ordering process and reducing the variation in supplies.

Various predictive analytics models are developed for application-specific functions [7].

Forecast Models One of the common predictive analysis models is forecast model. The new data is estimated based on the historical data. They are used to predict market trends. For example, "Internet of Medical Things, Forecast to 2021" [8] forecast the IoMT market in 2021, in hospital segment, on the body segment, etc.

Classification Model Classification is a predictive analysis model which learns from the training data set to understand characteristics/features of different classes. This trained model categories the new record based on the features. The accuracy of prediction of this method is better than the conventional methods and helps the medical practitioner in deciding the treatment plan. Some common classification techniques include decision tree, random forest, KNN and naïve Bayes.

Common example for classification is detecting the presence of abnormalities in the human body like cancer cells. The severity of the disease can also be accurately identified.

Clustering Clustering is grouping the data based on the similarities present in the data. Data with similar characteristics are grouped together. Clustering is an unsupervised ML algorithm. Clustering is helpful when complete knowledge of the samples are not available. One application of this model is grouping of patients with similar medical record.

Outliers Models Outlier model identifies the unusual data in the network. This can be used for identifying intrusion in the system.

Time Series Model Time is considered as one of the input to develop this model. This model predicts the trend over time period. Time series analysis was used to predict the trend of COVID-19 spread.

13.2.1.1 Steps in Predictive Analysis Model [7]

The following are the steps involved in modelling predictive analysis:

1. Determine the application which needs predictive analysis.
2. All ML algorithms are data-intensive. Relevant data is collected from interconnected IoMT devise.
3. The collected data needs to be cleaned and to be converted to acceptable format.
4. Create, test and validate the model.
5. Use the model for real-time problems.

13.2.2   Infrastructure for Data Analytics

As ML needs large database, hence the memory requirement and the computing capacity needed is more to extract meaningful information. Based on where the data analytics is implemented, the IoMT data analytics is classified into cloud computing, edge computing and fog computing [4]

Cloud Computing   The data generated through millions of smart IoMT devices are gathered and stored in the cloud which is a remote computing facility. Data processing using ML/DL is carried out in cloud. ML methods require large number of samples and more computing facility. Remote computing facility can satisfy the memory and computing requirement of ML algorithms.

Edge Computing   Edge computing is the setup of computational capability at IoMT network edge. Cloud computing handles massive data and is the centralized data processing facility where the overhead is high. The bandwidth and power requirement for sending data to the cloud is also more. Edge computing enables the data processing in the local network and hence decentralizes data processing. Edge computing resolves the single point failure problem. Instead of having a huge centralized server in cloud computing, small distributed servers can be used in edge computing. Cloud computing is useful for complex and delay-tolerant IoMT services. For real-time low latency operations, edge computing is the better choice.

Fog Computing   Fog computing is a compromise between cloud and edge computing. According to [4], in edge computing, data processing is carried out at the edge of the network. In fog computing the networking services are provided between the cloud and end devices. Fog computing provides processing and storage facilities through a virtual platform. Using cloud technologies data can be stored up to months or even years, whereas fog computing can store the data only for limited time.

Edge-cloud collaborative model is currently being used in IOMT networks. The devices of the network are connected to the cloud via a gateway where edge com-

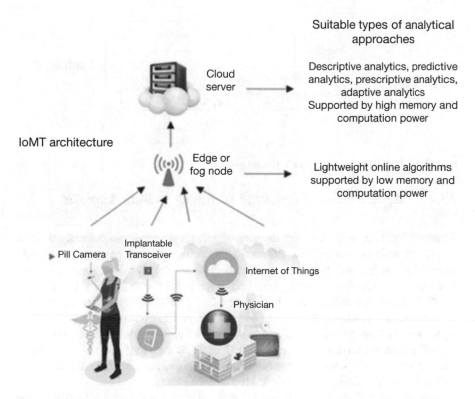

Fig. 13.2   Different IoMT architecture [4]

puting is carried out. Figure 13.2 describes different components in IoMT and the way it is connected to create the overall systems.

Data from IoMT should be processed to produce meaningful information. Integration of data analytics using ML in IoMT architecture is shown in Fig. 13.3 [4]. The end user of this in this architecture may be machine or human [4].

When there is communication between machines, it is necessary to have an application layer protocol to transmit data in the prescribed format to suit the lower layer protocols. There are many IoMT application layer protocols to meet the resource requirements of different IoMT devices. Requirements depend on the message size, overhead and delay. Constrained Application Protocol (CoAP) is less resource intensive and used in applications where the IoMT devices just senses the data and offload to the edge/cloud. For high-end IoMT devices, Message Queuing Telemetry Transport (MQTT) and Advanced Message Queuing Protocol (AMQP) are used. In this scenario data can be transmitted to other machines and IoMT devices. HyperText Transfer Protocol (HTTP) is more resource intense protocol and used in very high-end IoMT devices.

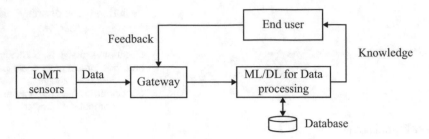

Fig. 13.3 ML/DL implementation for data processing

13.2.3 Applications of Machine Learning in Data Analysis

Research has been undertaken in the area of ML for data analysis for early disease detection, activity recognition, lifestyle monitoring, support patients to manage their daily routine, etc. [9]. One of the main biomedical signals to understand the body conditions is electrocardiogram (ECG). In [10], SVM classifier is used on the IoMT device to detect heartbeat abnormality. In [11], classification techniques were used to classify cardiovascular diseases using Bayesian network model. The ECG signal sensed from a wearable device is transmitted to a remote place, and Bayesian model is used to classify them. In [12] ECG signal is used to detect the stress level. Among the classification algorithms used, highest accuracy was obtained with Naive Bayes. The design was intended for IoMT devices.

Electroencephalogram (EEG) is another vital signal which can be analysed for the well-being of humans. Epileptic seizures is predicted [13] by classifying the EEG signal using logistic regression and gradient boosting algorithms. As LoR is a low overhead algorithm, it is deployed in IoMT device itself and the boosting algorithm in the gateway. The drowsiness of a driver is classified analysing the EEG signal using SVM implemented in the IoMT wearable device [14]. k-Means clustering algorithms implemented in gateway and fog node are used to analyse the physiological data collected by a wearable device [15].

Posture monitoring of patient based on the weight on specially designed matters using data collected with IoMT is researched [16] which helps in identifying the correlation with sleep pattern. Kidney ultrasound image collected using IoMT-enabled device was used with SVM to predict the abnormalities in the kidney [17].

According to the market research, big data and ML in healthcare industry are bound to grow in the future. IoMT-based solutions are making a big effect in the healthcare industry. The network of medical devices is expected to rise from $14.9 billion to $52.2 billion by 2022 [35]. Though there are tremendous advantages with IoMT, privacy and security of the data are the major concerns. As there are lot of devices connected through the Internet, it is vulnerable for various types of attacks. These attacks are serious concern as IoMT deals with the networking and controlling of smart medical devices. The existing security methods are not sufficient enough to handle this huge system with more interconnected devices. It is important to have an intelligent security system for IoMT environment.

13.3 ML/DL Methods for IOT Security

Billions of smart devices are connected together in IoMT to improve human life. Generally, IoMT devices are placed in diverse environments and provide interaction between their surroundings. This interconnection of devices makes the environment more vulnerable for different types of attacks. Both active and passive threats can affect the IOMT device. The existing security methods may not be sufficient, and it is very crucial to have an intelligent security system as it involves communication and controlling of devices which can affect the human life. ML/DL methods can be used to for secure communication between devices [18].

The important properties that need to be considered in IoMT security methods are listed below.

Confidentiality IoMT devices store and transfer highly sensitive and vital information. So unauthorized access of this data needs to be prevented.

Integrity Integrity ensures the modification of data only by authorized persons. Wireless transmission of IoMT data introduces integrity challenges. As IoMT deals with decision-making and treatment of human being, effective integrity checking is a must. Any error, loss of data or modification may have severe effects.

Authentication Before using the IoMT-generated data for any purpose, the identity of the data needs to be verified. As the authentication method for each system is different, connected robust authentication services need to be established. Security and safety need to be considered while designing the authentication schemes.

Authorization This deals with granting rights to access the IoMT system. Access should be granted to humans as well as sensors to interact with the IoMT system. Data should be made available only to the authorized entities. IoMT is a heterogeneous environment with lot of sensors interconnected; necessary check needs to be done to check the validity of the requester.

Availability One of the main features of IoMT is availability of services at anytime and anywhere. The security protocol should ensure the availability of IoMT services to the user uninterruptedly.

Non-repudiation This means providing access to actions that cannot be repeated again. In medical sector, non-repudiation is one of the important security properties.

An effective security scheme should consider all these properties. The security threats tries to break the security schemes in an IoMT system which can be categorized into cyber threats and physical threats [18].

13.3.1 Threats

Cyber Threats Passive threats: Passive threats do not have any harmful effect immediately. This is done by spying through the communication channel. By this an attacker will be able to collect valuable health information from different sensors and track the location of the sensors. This affects the privacy of the sensor holders.

Active threats: Active threats will have harmful effect in the overall system. The attacker in the active threats spies the communication channel and modifies the system configuration. The attacker can also control the communication through the channel, modify and disrupt the data that is communicated which may lead to serious problems. Some of the active threats may include impersonation, malicious inputs, data tampering and denial of services.

Variants of cyber threats are given in Fig. 13.4 [18].

Physical Threats As the devices of IoMT are available at multiple places, the attacker can damage any of these devices causing interruption in the service. Damage for the IoMT devices may happen due to natural calamities like earthquakes and floods.

An intruder may attack the IoMT services at different attack surface levels including physical device, network service layer, cloud service, web and application interface.

Security is a major concern in IoMT with increasing number of devices connected. Attacks on the network especially zero-day attacks are evolving in IOT and

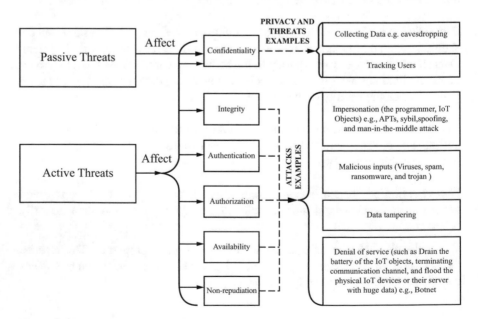

Fig. 13.4 Types of cyber threats in IoMT [18]

IoMT environments [18]. Zero-day attacks exploit the vulnerability in the system that hackers can use to attack the system. These attacks are very dangerous as they automatically reprogram themselves every time they are transmitted. Monitoring and controlling of these attacks are vital for the satisfactory operation of IoMT. Identifying these attacks with traditional security algorithms are very limited scope, and it is the need of the day to switch over to more intelligent and powerful tools to learn the normal and abnormal behaviour of the devices connected in IoMT environment.

13.3.2 Machine Learning (ML) Methods for IoT Security

For improving the security of IoMT devices, supervised and unsupervised learning algorithms are used [18] which are discussed.

Support Vector Machine SVM is the common classifier used, and it generates a hyperplane to separate between multiple classes. This is a maximum marginal classifier which means the hyperplane is constructed such that the distance between closest data point to the hyperplane is maximized for each class. SVM is a more generalized classifier and can class data set with higher number of features. Selection of optimal kernel for nonlinear class is a challenging task in implementing SVM. SVM is applicable to detect real-time intrusion [19]. SVM outperforms the ML algorithms including RF and NB in malware detection in IOT environment [20].

Decision Trees (DTs) Decision tree is classification method which divides the input into various regions based on the attributes of the input. Each node checks for an attribute and divides the input space into nonoverlapping subregions. Each leaf node belongs to one of the possible class. Optimal selection of parameters for breaking the tree at each level is carried out using different metrics [21]. The feature at each node is selected, such that the breaking of training set into subclass has minimum interaction between them. This improves the classifying efficiency of the classifier [21]. Decision tree is used for intrusion identification and also to detect suspicious source. Decision tree is simple when the nodes are limited in number. However, when the number of attributes and hence the number of nodes are more, DT needs more memory for implementation, and also it is complex to understand and trace.

Naive Bayes Classifier In statistical theory, Bayes' theorem estimates the probability of event happening based on the previous probability connected to that event. This theorem can estimate the probability of attack based on the previous traffic information. Naive Bayes (NB) classifier is a supervised classifier that uses Bayes theorem. The network attributes such as duration of the connection, protocol used for connection and status flag are used by Naive Bayes classifier. This classifies the traffic as usual or abnormal [22]. The simplicity, easy deployment and applicability to binary and multiple classes makes Naive Bayes classifier very attractive. The number of training samples needed for this classifier is also very less. Interaction between the features helps in improving the classification algorithm which is not possible with Naive Bayes classifier.

k-Nearest Neighbour (KNN) k-Nearest neighbour (KNN) is a nonparametric classification method. All the training samples are mapped on a P-dimensional plane, where P is the number of attributes/features of the input space. For new input sample, the distance from the new point to all the training samples is computed. Classification of the new sample point is done based on the votes of the nearest K neighbours. Euclidean distance is considered as the common distance measure. In security implementation, the unknown sample is classified as malicious or normal based on its behaviour. Though the algorithm is very simple, optimum value of K depends on the data set and application. Optimum value of K can be obtained by cross validation process by testing with different values of K. KNN is used for intrusion detection [23].

Random Forest (RF) Random forest belongs to supervised classification method. This is an extended version of decision tree classifier. RFs are supervised learning algorithms. Several randomly constructed decision trees are combined in RF to get a better classification accuracy. For a new sample input, each DT will select a class and the class which is voted by most of the DTs is selected as the class for the new input. Thus RF averages the output obtained by several randomly constructed DTs. This classification method also helps in intrusion and anomaly detection. As RF consists of many DTs, data set needed for construction is large and may not be implemented in real-time applications [24].

Association Rule (AR) AR algorithm learns the relationship between variables and understands the correlation between them and constructs a model based on that. This developed model can classify the new data sample. AR algorithms identify the variables which exist together in most of the attacks and used as a base for identifying the possible attacks [25]. In security applications, the association between the variables are not simple which in turn makes the algorithm more complex and also the number of association rules increases.

k-Means Clustering k-Means clustering belongs to unsupervised ML method. This is not a classification algorithm; rather this is a clustering algorithm. Unsupervised algorithms are used when there is no label data set available. k-Means clustering groups the points K clusters with cluster heads iteratively formed. Initially K cluster heads are selected, and the data are grouped. A data point is grouped under a cluster if the distance to that particular cluster head is minimum compared to other cluster heads. All the cluster head is recomputed based on the mean value of all the samples belonging to that cluster. These two steps are repeated several times to get a better result. Thus data is clustered iteratively depending on the distance between the data point and the cluster heads. Euclidian distance is the most common distance measures used. Each cluster will comprise data points with like features. Application of k-means clustering for IoMT security is not well developed [26], and the possibility of this has to be explored.

Ensemble Learning (EL) Ensemble learning combines the classification result of most of the above-discussed classifiers to arrive at a better classification result. Every classification methods discussed has some advantages and disadvantages. There is no classification method that will best suit for all the applications/data sets. The classification algorithms efficiency and performance vary with application [27]. The performance of a selected classifier also varies with the metric used (e.g., K value in KNN, etc.). Ensemble learning combines multiple homogeneous or heterogeneous classifiers to obtain better result. As EL combines many classifiers, the time complexity is more. For detecting intrusion, anomaly and malware EL-based classifier can be used.

Principal Component Analysis (PCA) The complexity of the ML algorithms increases based on the number of attributes/features of a data set. PCA is a feature-reduction technique used to reduce the number of attributes in the data set. The correlated features are identified and removed by PCA. The reduced set will have most of the information content of the original set and is called as principal component [28]. So PCA is used along with any classifier algorithm to get a less complex ML model. The ML methods for IOMT security is summarized in Table 13.1.

Recently, with the advances in deep learning techniques, it is being used in many areas where ML has limited use. Deep learning algorithms perform better than ML for applications with large data set. IoMT devices tend to produce bulk of data, and hence DL is more suitable with IoMT. DL will automatically learn suitable features from the data set. DL architecture has many processing levels called layers that learn many features with different abstraction levels. Some of the deep neural networks with their application in IoMT security is discussed.

Convolutional Neural Networks (CNNs) At the higher level, CNN consists of convolution layer, rectified linear unit (Relu) layer and max pooling layer. At the output side, CNN has fully connected layer, softmax layer and classification layer.

The convolutional layer of CNN comprises a number of filters. The weight of these filters decides the features that are extracted and is decided in the training phase. The basic features are learnt by the convolutional layer close to the input. These low-level features are combined and are recognized at the convolutional layer close to the output side. Max pooling layer is used to reduce the size of the features. This is also important to reduce the computational time. This layer keeps the value of the maximum pixel in the rectangular neighbourhood.

At the output side, fully connected layer are connected as feed forward multi-layer network. The number of connections in the fully connected layer is huge and hence the number of parameters to be learnt by the network increases. Depending on the output of these layer, the next layer which is softmax layer calculates the probabilities of a particular data belonging to different classes. Based on these prob-

Table 13.1 Machine learning methods for implementing IOMT security

Technique	Working principle	Advantages	Disadvantages
Support vector machine	Creates a hyperplane for classification	Can be used for data set with large number of features	Selection of kernel function is difficult in multi-class classification
Decision tree	Construct a decision tree based on the training samples	Simple and easy	Memory requirement is more and complexity increases with number of features
Naive Bayes classifier	Uses Bayes theorem for identifying the attacks	Simple to implement	The interaction between the features is not used to get better detection efficiency
k-Nearest neighbour	Classification is performed based on the class of the majority of k-neighbours	Simple and common ML method	Selection of optimal K value is challenging
Random forest	Combination of multiple DLs and the combined result is considered as the final result	Robust	Large data set is needed as many DLs need to be constructed
Association rule	Understands the relationship between different variables and a model is constructed based on that	Easy	Simple relationship between variables may not be correct in security-based applications
k-Means clustering	This is unsupervised learning. K clusters are created based on the features of the data. Data points with similar features are clustered iteratively.	This is particularly useful when labelled data set is not available	Less effective than supervised learning
Ensemble learning	Homogeneous and heterogeneous combination of ML models	Better performance and suitable for all type of data set	Complexity is more
Principle component analysis	Used along with other ML algorithms to reduce the number of features and hence the complexity. This is used before constructing the model. Only uncorrelated principle features are preserved and all the redundant features are removed	This simplifies ML algorithm by reducing the dimensionality	This is used along with other ML algorithms

abilities, the classification is done. CNN can be used for malware detection [29]. CNNs learn automatically from the raw security data and hence can be used in end to end security of IoMT system.

Recurrent Neural Networks (RNNs) Recurrent neural networks are also deep neural network which handles sequential data. RNN takes hidden states as inputs.

RNN has internals memory. After the output is produced for any input, it is stored in the recurrent layer of the network for further use. To produce any output, the input and the stored previous outputs are considered. In RNN the intermediate values are stored which contains information about the past inputs.

RNN can be very well used for improving the IoMT security. IoMT devices produce huge amount of data sequentially from different sources including traffic in the network which is one of the key parameters for sensing network attacks. In attack detection, prediction of pattern is time dependent. Therefore, recurrent connections will improve the prediction of important pattern behaviour. In [30], authors have explained the feasibility of using RNN for detecting potential attacks in IOT system by examining the network traffic. This model could detect the malicious attacks accurately. Variants of RNN can further be explored to implement IoMT security in an effective way.

Deep Autoencoders (AEs) Deep autoencoder belongs to unsupervised learning network which encodes the input such that the encoded signal will have optimum number of features which is called as code. The decoder function of the AE will reproduce the original data from the encoded data. AEs are used for detecting malware in the network [31].

Restricted Boltzmann Machines (RBMs) RBMs are an example for deep generative model. Generative models can learn the data distribution using unsupervised learning. The visible layer in RBM holds the input, and the hidden layers contain the latent data. In [32], anomaly detection was developed based on a discriminative RBM. Discriminative RBM combines different generative models for improving the detection accuracy. The results showed that the developed RBM worked well for trained data set. Further research should be done to make the model more efficient.

Deep Belief Networks (DBNs) DBNs also fall under generative methods [33]. Stacked RBM executing greedy layer-wise training is present in DBN. This is the reason for the robust performance of DBN in unsupervised situations. Malicious detection can be effectively identified using DBN.

Generative Adversarial Networks (GANs) GAN is recently being widely used DL model. GAN has generative and discriminative models. Generative model learns the distributed data and discriminative model predicts the probability of the data belonging to different distribution. GAN model can be used to evaluate the authentication of the sample data [34].

Ensemble of DL Networks (EDLNs) EDLN is nothing but the DL algorithms that works simultaneously for better result. This is EDLN. EDLN can handle data with high-dimensional and complex problems. Each individual algorithm homogeneous (the same family classifier) or heterogeneous (different family classifier) can be stacked to get better performance.

The resources such as memory and computational capacity are limited in IoMT devices which limits the deployment of ML/DL algorithms in the device level. Execution of ML/DL algorithms in the cloud reduces the computational requirements in the IoMT devices. However, this increases the wireless energy overhead.

With weak network connectivity, offloading the data to the cloud is limited and will not be available to the application. Edge computing/mobile GPU is another solution for implementing ML/DL algorithms to enhance IoMT safety. In Edge computing, a sever at the network end will perform the security check with minimum delay. The energy efficiency and the scalability are other advantages of edge computing. The network traffic load in such an environment is also reduced [18].

Integration blockchain with ML/DL for IoT security is another current area [35]. Blockchain uses decentralized architecture which overcomes the security challenges that are associated with centralized architecture. With decentralized architecture, the authentication of data depends on the approval of many entities. As IoMT devices are distributed, the digital ledger blockchain becomes significant in security issues. Machine becomes smarter with ML/DL. So both blockchain and ML methods will work together to enhance the security of IoMT. ML can assist blockchain in smarter decision-making. Blockchain is a decentralized database which enables data distribution between several nodes and hence generates large amount of data for the ML/DL. This improves the performance of ML/DL.

13.4   Conclusion

IoMT has made tremendous impact in the medical industry and is one of the fast-growing field in computing domain with an estimate of 50 billion connected devices by the end of 2020. Precise diagnosis of disease is crucial for successful and satisfactory patient care. ML algorithm plays a vital role in diagnosis, and lot of research is going on in this area. The performance of the ML algorithm depends on the data set. The continuous data generated by IoMT can help the ML algorithms to perform better. The first part of the chapter gives an overview of ML methods for data analysis in IoMT realm.

More number of connected devices in IoMT leads to high vulnerability in the healthcare industry. It is crucial for the healthcare systems to understand these vulnerabilities and provide security. Conventional security algorithms are inefficient to handle the increasing security issues. Machine intelligence can be leveraged to find the normal and abnormal behaviour of the IoMT devices. The second part of this chapter discusses the enhanced security methods using ML/DL algorithms. There are lot of potential challenges in this area which serves as a future research direction.

References

1. Joshi, R. R., & Mulay, P. (2020). Closeness Factor Based Clustering Algorithm (CFBA) and allied implementations—Proposed IoMT perspective. In *A handbook of internet of things in biomedical and cyber physical system* (pp. 191–215). Cham: Springer.
2. What is the Internet of Medical Things (IoMT)? – Mobius MD, www.mobius.md
3. Dey, N., Ashour, A. S., & Bhatt, C. (2017). Internet of things driven connected healthcare. In *Internet of things and big data technologies for next generation healthcare* (pp. 3–12). Cham: Springer.
4. Adi, E., Anwar, A., Baig, Z., & Zeadally, S. (2020). Machine learning and data analytics for the IoT. *Neural Computing and Applications*, 1–29.
5. Joyia, G. J., Liaqat, R. M., Farooq, A., & Rehman, S. (2017). Internet of Medical Things (IOMT): Applications, benefits and future challenges in healthcare domain. *Journal of Communications, 12*(4), 240–247.
6. Alharthi, H. (2018). Healthcare predictive analytics: An overview with a focus on Saudi Arabia. *Journal of Infection and Public Health, 11*(6), 749–756.
7. Types of predictive analytics models and how they work – Selerity, selerity.com
8. Internet of Medical Things, Forecast to 2021, store.frost.com
9. Samie, F., Bauer, L., & Henkel, J. (2019). From cloud down to things: An overview of machine learning in internet of things. *IEEE Internet of Things Journal, 6*(3), 4921–4934.
10. Azariadi, D., Tsoutsouras, V., Xydis, S., & Soudris, D. (2016, May). *ECG signal analysis and arrhythmia detection on IoT wearable medical devices*. In 2016 5th International conference on modern circuits and systems technologies (MOCAST) (pp. 1–4). IEEE.
11. Hadjem, M., Salem, O., & Naït-Abdesselam, F. (2014, October). *An ECG monitoring system for prediction of cardiac anomalies using WBAN*. In 2014 IEEE 16th International conference on e-Health networking, Applications and Services (Healthcom) (pp. 441–446). IEEE.
12. Keshan, N., Parimi, P. V., & Bichindaritz, I. (2015, October). *Machine learning for stress detection from ECG signals in automobile drivers*. In 2015 IEEE International conference on big data (Big Data) (pp. 2661–2669). IEEE.
13. Samie, F., Paul, S., Bauer, L., & Henkel, J. (2018, March). *Highly efficient and accurate seizure prediction on constrained iot devices*. In 2018 Design, Automation & Test in Europe Conference & Exhibition (DATE) (pp. 955–960). IEEE.
14. Li, G., Lee, B. L., & Chung, W. Y. (2015). Smartwatch-based wearable EEG system for driver drowsiness detection. *IEEE Sensors Journal, 15*(12), 7169–7180.
15. Borthakur, D., Dubey, H., Constant, N., Mahler, L., & Mankodiya, K. (2017, November). *Smart fog: Fog computing framework for unsupervised clustering analytics in wearable internet of things*. In 2017 IEEE Global Conference on Signal and Information Processing (GlobalSIP) (pp. 472–476). IEEE.
16. Matar, G., Lina, J. M., Carrier, J., Riley, A., & Kaddoum, G. (2016, September). *Internet of Things in sleep monitoring: An application for posture recognition using supervised learning*. In 2016 IEEE 18th International conference on e-Health networking, applications and services (Healthcom) (pp. 1–6). IEEE.
17. Krishna, K. D., Akkala, V., Bharath, R., Rajalakshmi, P., Mohammed, A. M., Merchant, S. N., & Desai, U. B. (2016). Computer aided abnormality detection for kidney on FPGA based IoT enabled portable ultrasound imaging system. *Irbm, 37*(4), 189–197.
18. Al-Garadi, M. A., Mohamed, A., Al-Ali, A., Du, X., Ali, I., & Guizani, M. (2020). A survey of machine and deep learning methods for Internet of Things (IoT) security. *IEEE Communications Surveys & Tutorials*.
19. Liu, Y., & Pi, D. (2017). A novel kernel SVM algorithm with game theory for network intrusion detection. *KSII Transactions on Internet & Information Systems, 11*(8).

20. Ham, H. S., Kim, H. H., Kim, M. S., & Choi, M. J. (2014). Linear SVM-based android malware detection for reliable IoT services. *Journal of Applied Mathematics, 2014*.
21. Kotsiantis, S. B. (2013). Decision trees: A recent overview. *Artificial Intelligence Review, 39*(4), 261–283.
22. Mukherjee, S., & Sharma, N. (2012). Intrusion detection using naive Bayes classifier with feature reduction. *Procedia Technology, 4*, 119–128.
23. Adetunmbi, A. O., Falaki, S. O., Adewale, O. S., & Alese, B. K. (2008). Network intrusion detection based on rough set and k-nearest neighbour. *International Journal of Computing and ICT Research, 2*(1), 60–66.
24. Chang, Y., Li, W., & Yang, Z. (2017, July). *Network intrusion detection based on random forest and support vector machine*. In 2017 IEEE international conference on computational science and engineering (CSE) and IEEE international conference on embedded and ubiquitous computing (EUC) (Vol. 1, pp. 635–638). IEEE.
25. Tajbakhsh, A., Rahmati, M., & Mirzaei, A. (2009). Intrusion detection using fuzzy association rules. *Applied Soft Computing, 9*(2), 462–469.
26. Xie, M., Huang, M., Bai, Y., & Hu, Z. (2017). The anonymization protection algorithm based on fuzzy clustering for the ego of data in the internet of things. *Journal of Electrical and Computer Engineering, 2017*.
27. Aburomman, A. A., & Reaz, M. B. I. (2016). A novel SVM-kNN-PSO ensemble method for intrusion detection system. *Applied Soft Computing, 38*, 360–372.
28. Zhao, S., Li, W., Zia, T., & Zomaya, A. Y. (2017, November). *A dimension reduction model and classifier for anomaly-based intrusion detection in internet of things*. In 2017 IEEE 15th Intl Conf on Dependable, Autonomic and Secure Computing, 15th International Conference on Pervasive Intelligence and Computing, 3rd International Conference on Big Data Intelligence and Computing and Cyber Science and Technology Congress (DASC/PiCom/DataCom/CyberSciTech) (pp. 836–843). IEEE.
29. McLaughlin, N., Martinez del Rincon, J., Kang, B., Yerima, S., Miller, P., Sezer, S., ... & Joon Ahn, G. (2017, March). *Deep android malware detection*. In Proceedings of the Seventh ACM on Conference on Data and Application Security and Privacy (pp. 301-308).
30. Torres, P., Catania, C., Garcia, S., & Garino, C. G. (2016, June). An analysis of recurrent neural networks for botnet detection behavior. In 2016 IEEE biennial congress of Argentina (ARGENCON) (pp. 1–6). IEEE.
31. Yousefi-Azar, M., Varadharajan, V., Hamey, L., & Tupakula, U. (2017, May). *Autoencoder-based feature learning for cyber security applications*. In 2017 International joint conference on neural networks (IJCNN) (pp. 3854–3861). IEEE.
32. Fiore, U., Palmieri, F., Castiglione, A., & De Santis, A. (2013). Network anomaly detection with the restricted Boltzmann machine. *Neurocomputing, 122*, 13–23.
33. Hinton, G. E., Osindero, S., & Teh, Y. W. (2006). A fast learning algorithm for deep belief nets. *Neural Computation, 18*(7), 1527–1554.
34. Hiromoto, R. E., Haney, M., & Vakanski, A. (2017, September). *A secure architecture for IoT with supply chain risk management*. In 2017 9th IEEE International Conference on Intelligent Data Acquisition and Advanced Computing Systems: Technology and Applications (IDAACS) (Vol. 1, pp. 431–435). IEEE.
35. Dilawar, N., Rizwan, M., Ahmad, F., & Akram, S. (2019). Blockchain: securing Internet of Medical Things (IoMT). *International Journal of Advance Computer Science Applications, 10*(1).

Index